PRAISE FOR *THE LAST STARGAZERS*

"Through captivating stories, Levesque gives us both a vivid and accessible inside look at the enigmatic mountaintop astronomers. A unique and engaging read."

—Sara Seager, professor of planetary
science and physics at MIT

"Emily Levesque is smart and funny, and her insider's tale of stars and the astronomers who study them bursts with color and energy."

—Edward Dolnick, author of *The Clockwork Universe*

"Emily's book is a compulsive read. It demonstrates what being an observational astronomer is really like—the highs, the lows and the unscheduled things that can happen at telescopes around the world! Give this book to every young person (especially the girls!) that you know who likes math and science."

—Jocelyn Bell Burnell, astrophysicist, Oxford

"Astronomy is *dangerous*. Wild (sometimes venomous) animals, thin air, heavy equipment, hazardous chemicals… Dr. Levesque captures all this with amusement and personal experience, making this a delightful read for everyone."

—Phil Plait, astronomer and author of *Bad Astronomy*

"It's like catching a glimpse of the magic behind the curtain galaxies away and leaves you hanging on every spectacular word. A must-read for anyone who has looked up at the sky and felt a sense of wonder as well as those considering the world of astrophysics and astronomy."

—Tamara Robertson, host of *Mythbusters:
The Search* and STEM speaker

"If you've ever wondered what astronomers do—what they *really* do—and the human journey from the era of eyepieces to gigantic robotic cameras, *The Last Stargazers* puts you there with compelling honesty, following the scientists and students with hundred-ton telescopes as backdrop."

—Erik Asphaug, author of *When the Earth Had Two Moons*

"Levesque, a University of Washington astronomy professor, leads readers on a pilgrimage to observatories throughout the world in her wonderful debut. This will particularly appeal to young women interested in science, but any stargazer would enjoy this joyous adventure through modern astronomy."

—*Publishers Weekly*, Starred Review

"Entertaining, ardent tales from an era of stargazing that may not last much longer."

—*Kirkus Reviews*

THE **LAST STARGAZERS**

THE MILKY WAY GALAXY

PERSEUS

SCUTUM-CENTAURUS

NORMA

SOLAR ORBIT

CARINA-SAGITTARIUS

ORION-CYGNUS

THE **LAST STARGAZERS**

THE LAST STARGAZERS

THE ENDURING STORY OF ASTRONOMY'S VANISHING EXPLORERS

EMILY LEVESQUE

Published by Sourcebooks
P.O. Box 4410, Naperville, Illinois 60567-4410
(630) 961-3900
sourcebooks.com

Library of Congress Cataloging-in-Publication Data

Names: Levesque, Emily, author.
Title: The last stargazers : the enduring story of astronomy's vanishing
   explorers / Emily Levesque.
Description: Naperville, Illinois : Sourcebooks, [2020] | Includes
   bibliographical references and index.
Identifiers: LCCN 2020005285 | (hardcover)
Subjects: LCSH: Levesque, Emily. | Astronomy--Popular works. |
   Astronomers--Popular works. | Astronomical observatories--Popular works.
Classification: LCC QB44.3 .L45 2020 | DDC 520--dc23
LC record available at https://lccn.loc.gov/2020005285

Printed and bound in the United States of America.
MA 10 9 8 7 6 5 4 3 2 1

For Mom,
who gave me stories

CONTENTS

INTRODUCTION

"Have you tried turning it off and back on again?"

This phrase, repeated by weary IT specialists the world over, had possibly never prompted such horror. First of all, it was one in the morning, and I was sitting in a chilly control room on top of the highest mountain in Hawaii. I was nearly fourteen thousand feet above sea level, twenty-four years old, and desperately fighting through sleep and oxygen deprivation to salvage several hard-won hours of PhD thesis research time on a piece of broken equipment.

Second, the equipment in question was the Subaru Telescope, a 630-ton beast housed one floor above my head in a fourteen-story dome. A joint venture between the U.S. and Japanese astronomy communities, the telescope boasted a pristine primary mirror more than twenty-seven feet in diameter (the largest single piece of glass in the world) and a suite of some of the most sophisticated scientific instruments and imaging tools on the planet. It cost $47,000 per night to operate, and after submitting a twelve-page science proposal to the professors in my department, I had been granted one of these valuable nights—tonight, the only night allotted to me in the entire year—to point this telescope at a handful of galaxies five billion light-years away.

No, I had not tried turning it off and back on again.

The evening had been going excellently until one of the control room computers had produced an unsettling *bloonk* sound and prompted the telescope operator—the only other person with me on the mountain—to freeze in her seat. When I asked what was up, she cautiously informed me that one of the mechanized supports holding up a mirror had just failed, but "it's okay. I think the mirror is still on the telescope."

"You *think*?"

"Yeah. If it wasn't, we would have heard a crash." Solid reasoning, if not exactly reassuring.

We'd apparently gotten lucky with how the telescope was positioned when the mechanized support failed, preventing an immediate disaster. For the moment, it was still holding up the secondary mirror, notably smaller than the primary but still four feet wide, four hundred pounds, and suspended seventy-three feet in the air to redirect light collected and reflected by the primary mirror into the camera I was using. Unfortunately, if we moved the thing again, we'd be at risk of dumping the secondary onto the floor, and that was if we were lucky. If we were unlucky, it would hit the primary on its way down.

We put in a nervous call to the Subaru members of the day crew, a group of engineers who worked on maintaining the thirteen telescopes on the mountain during the daylight hours when the observers were asleep. The Japanese crew member we reached cheerfully informed us that he had, in fact, seen this happen earlier in the day, that the mechanized supports were *probably* fine and it was *probably* just a false alarm, and that turning the power off and then on again would *probably* fix the problem. It seemed impolite to point out that we were talking about a multimillion-dollar telescope and not a modem.

I didn't know what four hundred pounds of glass hitting the concrete floor above my head would sound like, but I knew I didn't want to find out. I was also quite sure I didn't want to be forever known as "the grad student who killed Subaru." I'd heard too many "I broke the telescope" stories over the years to ignore the fact that this was a very real possibility. One of my collaborators had destroyed an outlandishly expensive digital camera on a telescope by innocently touching two of the wrong wires together; the story had gotten back to his boss before he did. Another astronomer, this one a veteran observer, had slammed the business end of a telescope into a movable platform inside the dome that she had forgotten to retract partway through a sleep-deprived night. Sometimes these sorts of failures weren't even anyone's fault. A gargantuan three-hundred-foot-wide radio telescope in Green Bank, West Virginia, had just up and collapsed one evening, crumpling like a stepped-on soda can partway through an observation. I couldn't remember exactly what had caused the infamous Green Bank failure, but I was convinced the words *mechanized support* had been involved. The cautious thing for me to do would be to call it a night, drive back down to the observatory's sleeping quarters, and have the day crew carefully check things over the next morning.

On the other hand, this was my *only* night on the telescope. Tomorrow, it wouldn't matter whether I'd experienced a mechanical failure, a false alarm, or even just some poorly timed clouds; telescope time is strictly scheduled months in advance, and another astronomer would be arriving with a completely different science program. All that would matter was that my night had come and gone without completing my observations. I would have to submit a whole new proposal, hope for another hard-to-get yes from the telescope committee, wait an entire year—a full trip of the earth around the sun—until the galaxies were back up in the night sky to try again, and hope *that* night wouldn't have any clouds or telescope problems.

I desperately needed these galaxies. Several billion years ago, each of them had hosted a strange phenomenon known as a gamma-ray burst. Astronomers' best guess was that these bursts were coming from massive rapidly spinning dying stars whose cores were collapsing into black holes and cannibalizing the stars from the inside out, igniting violent jets of light that came streaking through the cosmos to arrive at Earth as flashes of gamma rays lasting mere seconds. Stars died all the time, of course, but only a handful of them were flashing us like this, and nobody could explain why. I had built my entire PhD thesis on the idea that studying the chemical makeup of these stars' home galaxies—the same gas and dust they had been born from—was the key to understanding why they exploded the way they did. Subaru was one of the only telescopes in the world capable of such observations, and the day crew had said it was *probably* giving me a false alarm. If I called off the night, I'd be giving up what could well be my only opportunity to ever study these galaxies, losing a linchpin of my thesis research in the process.

Of course, having the largest piece of glass in the world sitting in pieces on the dome floor wouldn't help matters either.

I looked at the operator, and she looked back at me. I was the astronomer, so with all of my twenty-four-year-old, third-year-grad-student, still-had-to-pay-the-young-driver-fee-to-rent-a-car wisdom, this was my call. I looked at the printout of my meticulously crafted observing plan for the night, which was falling further and further behind with every minute that Subaru sat idle. I looked at the fuzzy image of the night sky on my computer screen, coming from the small guider camera that was always on and showing us where the telescope was pointed to help observers like me find their way through a bottomless sea of stars.

I turned the power off and back on again.

○ ○ ○

The simple act of stargazing is an experience shared by almost every human on the planet. Whether we're peering through the stifling light pollution of a bustling city, struck motionless by the riot of stars arcing over our heads in a remote corner of the globe, or simply standing still and feeling the enormity of space waiting just outside our planet's atmosphere, the beauty and mystery of the night sky has always entranced us. You'd also be hard-pressed to find someone who hasn't admired the dramatic astronomical photographs produced by the world's best telescopes: the sweeping vistas of stars, galaxy pinwheels, and rainbow-hued gas clouds that supposedly hold the secrets of the cosmos.

What's less well-known is the story behind *where* these photos come from, how and why we're taking them, and who exactly is extracting those secrets of the universe. *Astronomer* sounds like a romantic and dewy-eyed sort of job, and its practitioners are a unicornesque rarity: of the 7.5 billion people on our planet, fewer than fifty thousand are professional astronomers. Most people have never even *met* a professional astronomer, let alone contemplated the details of such a strange career. When thinking about what an astronomer does (on the rare occasion that it's thought of at all), people tend to imagine their own experiences with stargazing taken to an obsessive level: a nocturnal geek peering through a really big telescope in a really dark place, maybe donning a white lab coat and confidently spouting off the names and locations of things in the sky as they patiently sit in the cold and wait for their next discovery. The handful of astronomers in movies also become a go-to reference: Jodie Foster hunkering down with headphones to listen for aliens in *Contact* or Elijah Wood peering through a suspiciously powerful backyard telescope to discover a planet-destroying asteroid in *Deep Impact*. In almost every case, the observing is merely the prelude to the real drama; the sky is always clear, the telescope is always working, and after a minute or so of wide-eyed awe, the movie astronomer dashes off to save the world with a few snippets of perfect data in hand.

This was certainly the mental image of astronomy that I had in mind when I claimed it as my future career. I'd come to astronomy in the same way as countless other amateur and professional space enthusiasts, through a childhood of backyard stargazing in a New England factory town, Carl Sagan's writing on my parents' bookshelf, and those jaw-dropping photographs of nebulae and star fields that seemed to always show up as the backdrops of TV specials and science magazine covers. Even when I arrived at MIT as a freshman and blithely declared myself a physics major in my first step toward an astronomy career, I had only a vague sense of what I'd be doing all day in my chosen profession. I was becoming an astronomer because I wanted to explore the universe and learn the stories of the night sky; beyond those broad strokes, I wasn't particularly fussed with what the exact job description of "astrophysicist" entailed. My daydreams were about contacting aliens, unraveling the mysteries of black holes, and discovering a new type of star. (So far, only one of these has come true.)

I did not daydream about being the final decision point for keeping one of the world's largest telescopes intact. I never imagined that one day, I'd be shimmying up the support struts of a different telescope to duct-tape a piece of foam across its mirror in the name of science, researching whether my employer carried experimental aircraft insurance, or willing myself to somehow fall asleep next to a tarantula the size of my head. I didn't know that there were astronomers who traveled to the stratosphere and the South Pole for their work, astronomers who had braved polar bears, faced down gunmen, and even died in pursuit of a few precious bits of light.

I also had no idea that the field I was entering was changing as rapidly as the rest of the world. The astronomers I read about and imagined—swathed in fleece, perched behind an impossibly large telescope on a cold mountaintop and squinting into an eyepiece while the stars wheeled above them—were already an endangered and evolving species. In joining their ranks, I would fall even deeper in love with the beauty of space, but to my surprise, I would also wind up exploring my own planet and learning the stories of an incredible, rare, and rapidly changing—even vanishing—field.

CHAPTER ONE

FIRST LIGHT

TUCSON, ARIZONA
May 2004

I got my first glimpse of a telescope—a real, large, world-class-observatory telescope—on the road heading west from Tucson. I had just finished my sophomore year at MIT, flying to Arizona straight from my final exams in quantum physics and thermodynamics, and was picked up at the Tucson airport by Phil Massey, an astronomer with gray mad-scientist curls, black-rimmed glasses, and a wide grin. My research adviser for the next ten weeks, he was driving me to Kitt Peak National Observatory, deep in the Sonoran Desert, where we'd be spending five nights observing at one of the telescopes as a kickoff to my summer project. It would be my very first visit to a professional observatory.

I had learned from our email exchanges that I would be studying red supergiants. Red supergiants are massive stars with at least eight times as much mass as our own sun. Because of their large masses, they've sped through their stellar lives at breakneck speed, taking a mere ten million years to transform from their newborn state—brilliant blue-hot stars freshly formed out of gas and dust—to their current state, blazing deep red like a dying ember and swelling up to many times their original size in a last-ditch effort to stay stable and alive. Death for these stars most likely means a violent interior collapse followed by a rebound explosion known as a supernova, one of the most luminous and energetic phenomena in the universe and the process by which black holes are sometimes formed.

Phil and I had met just once before, briefly, the previous January, and he'd chosen me as his summer student based on a presentation he'd seen me give on my first foray into astronomy research. When we'd first started discussing plans for the summer, Phil had offered me a choice of two projects, literally red or blue: the red dying stars or the blue newborn ones. I didn't know a great deal about either, but I thought black holes were fascinating, and since the dying stars seemed fractionally closer to that point, I opted for red. At Kitt Peak, Phil and I would be observing about a hundred red supergiants in our own galaxy, the Milky Way. I'd then spend the rest of the summer working with the resulting data, trying to measure the stars' temperatures and contribute a tiny piece to the ongoing astronomy-wide puzzle of exactly how these stars evolved and died.

Phil and I chatted and got to know each other on the drive, but I was also gawking out the window at the southern Arizona desert. The baking summer heat and sunlight were stunning, a world away from the muggy green spring I had left behind in Massachusetts, and I took in the orange-brown dirt and stretches of saguaro cactus and beaming blue sky. Phil pointed out the tiny white silhouette and pair of contrails from a high-altitude jet and mentioned that experienced astronomers can gauge the quality of the sky they'll be observing that night based on how long the contrails are. If they're long and fluffy, there's a lot of moisture in the atmosphere to stir up and interfere with the starlight, while if they're short—just a little tuft trailing behind the plane—we would be in for a crisp and clear night. The jet we were watching had a short tail.

Phil knew the drive to the observatory by heart and told me where to look at the exact moment when Kitt Peak's four-meter telescope first popped into view. The white dome, eighteen stories tall, glinted in the pounding desert sun. The telescope inside has made groundbreaking observations of everything from nearby stars to impossibly distant galaxies in the decades since it achieved "first light"—the moment when the completed telescope took its first look at the night sky—in 1973.

The vast majority of modern-day telescopes use mirrors to collect light from the stars, and the most fundamental property of these telescopes is the mirror's size. A larger mirror means that when we point the telescope at an object, a bigger area is available to collect light from the object. (It's the same principle behind why your pupils dilate in a dark room.) The distance from one side of a mirror to the other—its diameter—also dictates how sharp of an image the telescope can produce, like using a telephoto lens to get a clear photo of something small and far away. For well over a century, major strides in astronomy have revolved around the progression toward bigger and bigger mirrors, with their diameters dictating a telescope's fundamental ability to see farther into space. As a result, mirror size has become the defining characteristic of a telescope, to the point where it's sometimes entwined in the telescope's name or even defines the name entirely. At Kitt Peak, the flagship telescope is widely referred to as "the four-meter."

Eventually, we peeled off Route 86—already an incredibly barren and empty stretch of highway—and started winding our way up a meandering mountain road. At first, there was very little sign that we were going anywhere but deeper into the desert: long stretches of pavement, some switchbacks, and minimal signs of any life at all beyond the cacti. The only clue that we were heading to an observatory was the occasional curve of a white dome peeking out from between the hills. Later on, we started getting a few hints that we were not on just any mountain. As we neared the summit, signs started appearing, imploring nighttime drivers not to use high beams and eventually not to use any headlights at all in an effort to preserve the mountain's darkness.

Today's best observatories are built in the high, dry, remote places of the world. High altitudes give us a slightly thinner atmosphere and less turbulence in the air between the summit and the stars. Deserts mean air devoid of water vapor and moisture, good for weather and for image quality. The reasoning behind the remote locales is a bit more obvious: the farther we are from the rest of the world, the darker the skies (although even the darkest parts of the planet are fighting a constant battle against encroaching light pollution).

Kitt Peak lies near the southern border of the United States, less than thirty miles from the Mexican border. The mountain itself is all brown rock and stubby trees, indistinguishable from the desert around it but for two things: the white domes hunkered like sleeping giants across the long summit ridge and the invisible but very real perfection of the air passing over the summit. The land around the observatory largely belongs to the Tohono O'odham Nation. A prominent rock formation in the distance, shaped surprisingly like a telescope dome, is known by them as Baboquivari and is, in their cosmology, the center of the universe.

As our car climbed, I found myself wondering: *what's a professional observatory going to be like?* I had a mental picture of a big behemoth of a telescope like the one we'd spotted from the road, perched white and alone on some stark rock outcropping of a mountain ridge, but that was about it. I hadn't much thought about details like where we'd sleep (*during the day? would we sleep?*), what we'd eat (*should I have brought some snacks?*), or any of the other logistics. I figured it would all sort itself out and focused instead on drinking in our surroundings as we neared the summit.

o o o

TAUNTON, MASSACHUSETTS
1986

Not quite knowing what was coming was not a new sensation. I'd long since come to terms with the idea of optimistically plowing forward with "I want to be an astronomer!" as my guiding plan.

I'd been enraptured by space for as long as I could remember, but the original spark could be traced back to early 1986, when Halley's Comet made its most recent close flyby of the earth. My parents and older brother and I lived in a suburb of Taunton, Massachusetts. A blue-collar southern New England city with industrial roots, it nevertheless gave way to forested streets and ponds and cranberry bogs once you made it a few highway exits out of town—dark enough for stargazing.

Neither of my parents were scientists by training. Before I was born, they'd both earned teaching degrees, with a focus in special education. My mom worked as a speech therapist but eventually went back to school for a graduate degree in library science and moved up through the library positions of the Taunton school system. My dad had studied teaching but worked as an independent truck driver for years while becoming a self-taught computer expert, and by the time I was born, he was working at an insurance company as an IT specialist.

Still, both of them were scientists by nature, fundamentally curious about the world around them and constantly eager to learn as much as they could about whatever corner of it might catch their attention. My dad had taken an astronomy class as an elective in college at Northeastern, which made enough of an impression for him to take it up as an interest and pass the enthusiasm along to my mom.

In keeping with their lifelong habits, once my parents got into something, they were *into* it, full bore. When that something was astronomy, my dad scraped up funds to buy a backyard Celestron C8, a squat orange cylinder with an eight-inch mirror, and built his own table to mount it on along with some added shelving to store eyepieces, equipment, and a copy of *Norton's Star Atlas*. The flames were further fanned by Carl Sagan's *Cosmos* television series premiering in 1980, which prompted my librarian mom to stock up on his books. By the time I was born in 1984, astronomy was a background buzz in our house in the same vein as gardening, woodworking, birds, and classical music. My parents were determined to give my brother and me a rich and varied set of potential interests to explore.

Still, the real catalyst sparking my interest in astronomy was my brother, Ben, almost ten years my senior. I'm fairly convinced that when two siblings are this far apart in age, hero worship is just part of the package. Growing up, Ben was always the first and foremost arbiter of all things cool in my eyes, and he was endlessly patient with me rather than annoyed by a tiny tagalong. Ben played the violin, and therefore I asked to play the violin. Ben did science fair projects, so I started fashioning nonsensical "experiments" out of whatever toys or household items I could get my hands on. I even wanted braces because Ben had them (an opinion that was reversed rapidly once I was the one in the orthodontist's chair).

In February 1986, I was eighteen months old, and Ben was eleven, studying Halley's Comet for a school project. These sorts of projects always became full-family endeavors, so all four of us tromped out into the backyard one cold winter night, armed with our eight-inch telescope and its homemade table, to get a glimpse of this once-in-a-lifetime comet flyby (it's next due back in 2061). According to my parents, I was brought out to get a brief look with the worry that I might be a typical fussy toddler, scared of the dark and eager to get back inside. Instead, I was entranced: gaping up at the sky, staring through the telescope (in retrospect, I'm amazed that a not-yet-two-year-old could look through an eyepiece, but they swear I managed it), and refusing to go back inside as long as Ben was still observing.

The love of astronomy stuck in a way that my love of braces didn't. I was an early and voracious reader, and a few years after Halley's Comet,

I was learning about star clusters and black holes and the speed of light thanks to Geoffrey T. Williams's Planetron books, which chronicle the adventures of a little boy with a toy that transforms into a magical spaceship and sweeps him off to explore the heavens. I have a strong memory of being five, reading about how fast the speed of light was, and repeatedly flicking the light switch on and off in my room to convince myself that yep, once I flipped it on, the light arrived pretty much instantly. That seemed pretty fast to me.

Later, I inhaled every astronomy book I could get my hands on, watched Mr. Wizard and Bill Nye on TV, and went to every movie about scientists and space that came along. I remember particularly enjoying the movie *Twister* because it gave me an encouraging look at what scientists themselves might actually be like. The fictional tornado researchers on screen were doing cool and exciting research and having fun along the way, and the main character was a woman who rolled around in the mud and was obsessed with science but still managed to end the movie with a great kiss (a combination I'd already been warned might not be tenable in the long run thanks to plenty of other movies featuring women who Had to Choose between Careers and Men).

My parents did what they could to encourage my interest in space, but opportunities to explore a career in astronomy aren't exactly found on every street corner. None of us even

Age six, sporting my beloved new Hubble Space Telescope T-shirt shortly after its 1990 launch. *Credit: Henri Levesque.*

knew a professional scientist, let alone an astronomer, and while my entire extended family was filled with kind and bright and enthusiastic people, nobody had a PhD or knew much about what this sort of job entailed. All four of my grandparents had left school at young ages, despite being uniformly strong and passionate students, to work in local factories and contribute income to their families. My maternal grandmother in particular had been devastated by this and wept the day she left school; she later returned to complete a high school degree alongside my grandfather, Pépère, and went on to get a practical nursing degree while raising five kids, with Pépère working at the big silver factory in town. My parents and some of my aunts and uncles had all been first-generation college students, drinking in as much education as they could but ultimately getting practical degrees that would lead to good jobs: engineering, actuarial science, teaching. It was a big, loud, and exceedingly loving family, buoyed by an immense amount of collective curiosity and a love of learning for learning's sake, but nobody had a road map at hand for how to get started on a career in something as intangible and fanciful as astronomy.

I did get to chat with a professional astronomer once during my childhood. Our house was a twenty-minute drive away from Wheaton College, a tiny but excellent liberal arts college. When I was seven, my parents took me to a public stargazing night at the campus rooftop observatory, and I quickly informed the professor running the event that I wanted to be an astronomer. He bent down to my height, looked me right in the eye, and said, "Take as much mathematics as you can." I stared seriously back at him and responded, "Okay." From then on, math became my focal subject in school. I skipped one grade of math, then another, falling into complex bus arrangements for a few years to get me between the high school where I took freshman geometry and the middle school where I took seventh grade everything else.

In July 1994, a flurry of astronomical excitement hit the news when word got out that the Shoemaker-Levy 9 comet was on a collision course with Jupiter. As the strike approached, speculation both inside and out-side the astronomy community was focused around what would happen to Jupiter after being hit by a comet. Would we see any signs of the impact? The fabulous new Hubble Space Telescope was scheduled to observe it, but nobody was quite sure what they would see.

After the impact, the news came quickly that the view exceeded all expectations. The comet strike had left what looked like a spray of stark dark-brown bruises across Jupiter's lower flank. I remember a clip being shown over and over of a group of astronomers huddled around a few computer monitors at the Space Telescope Science Institute in Baltimore, grinning and gasping with excitement at what they could see. At the heart of the group, a bespectacled young woman named Heidi Hammel sat front and center, gleefully celebrating with her companions as spectacu-lar images of Jupiter came rolling in. My dad and I brought the backyard telescope outside soon after and spotted the impact scars on Jupiter with our own eyes, but that glimpse of the excited astronomers was what made a lasting impression on me. These were people who loved astronomy just as much as I did, doing this for their *job*, and feeding off one another's excitement. That could be me.

This particular moment stuck with me because despite having a vibrant and supportive family and happily immersing myself in science, I was also frequently a lonely and frustrated kid. I was the only kid in my school who liked astronomy instead of Nickelodeon, who took violin les-sons instead of ballet and soccer, and who was skipping around schools as I blasted through math classes. I was well aware that I was the weird kid who listened to classical music on her Walkman, who watched documentaries about squid instead of popular TV shows and movies, and who preferred beat-up cargo pants and T-shirts with math jokes on them to whatever was trendy at the time. I chafed against the isolation; I *wanted* friends, loved the idea of playing with other kids and having adventures and messing with sparkly nail polish and platform flip-flops (it was the 1990s), but not enough to give up who I was. I wanted friends who shared my love of space and math and old musicals. I wanted to be a world-famous astrophysicist, the first woman on Mars, the next Carl Sagan, but I also wanted to go on dates and get kissed and share all my imagined adventures with someone. I refused to believe that this was an impossible proposition. Surely, I couldn't be the *only* kid like me in the world.

I had hope thanks to the summer geek camps I was able to attend. In seventh grade, I scored high enough on the SAT to gain entrance to the Johns Hopkins Center for Talented Youth summer program. There I had my first encounter with other kids like me, a group who thought being able to play a Mozart violin concerto and do trigonometry made me *cool*, not a social pariah. I took an astronomy class through the program during the summer after ninth grade, and it blew my mind. I'd gone from being the only kid I knew who liked astronomy to being surrounded by an entire *class* of them. My people, it seemed, were out there—I just had to get to them. It was these summer experiences, along with the science fairs, music lessons, and a whole lot of time studying for AP exams, that launched me toward college at the ultimate geek mecca.

Most of my extended family was in the room when I found out I'd been accepted to MIT. The whole lot of us had just gotten back from an afternoon in Boston: my cousin Nathan and I had qualified to perform at the all-state music festival as a saxophonist and violinist, respectively, and in keeping with family habit, about twenty people had invaded Symphony Hall to watch us perform and then piled back into my parents' house for pizza and postconcert celebrating. In the midst of all this, I'd changed out of my concert clothes and padded barefoot down the driveway to grab the mail. I'd been deferred in MIT's early action admissions round and had all but written them off, so a giant fat envelope in the mailbox from them didn't even register as a blip. I lived in Massachusetts, after all, and it wasn't unusual to get pamphlets from MIT advertising this or that engineering program.

I toted the mail back into the house, where a few surprised looks at the envelope convinced me to open it in the kitchen as everyone milled around. When a folder slid out with CONGRATULATIONS CLASS OF 2006 emblazoned on it, I gawked, dumbfounded, as the entire room erupted around me. My parents and brother were delighted, and my cousins and aunts and uncles all cheered raucously. Meanwhile, my grandfather— Pépère, the undisputed heart and anchor of the family—leaned back and hitched up his belt, the same tell he used when he had a killer hand in cards, as a slow, wide smile took over his face. He'd always been convinced that my cousins and I were each geniuses on our way to changing the world, so he alone seemed unsurprised by the envelope still sitting in my hands. It was one of those rare moments where I knew *then*, not just in retrospect, that I was seeing my life change tracks in front of me.

My family met my declaration that I was going to study physics and become a professional astronomer with an attitude I can best describe as supportive trepidation. Lots of "That's awesome! You go for it! Exactly what kind of job will you get with that?" Apparently behind the scenes, there were lots of murmured conversations about the practicality of my majoring in something as abstract as physics. Things like engineering or biology at least had obvious endpoints and employment options, but none of us, myself included, had much idea of where the career path for a physicist, let alone an *astrophysicist*, would head. It was my brother, Ben, who eventually settled matters by pointing out that since this story would still end with me holding a physics degree from MIT, odds were I could probably convince someone to hire me.

Now I just had to make it out with that degree and, ideally, figure out what astronomers actually *did* along the way.

o o o

KITT PEAK NATIONAL OBSERVATORY, ARIZONA
May 2004

Once Phil and I arrived at the Kitt Peak summit, there was a quick flurry of checking in and finding our assigned spartan but comfortable rooms in the observatory dormitory before I was taken on a quick tour of the mountain. Our first stop was the towering four-meter telescope I'd been admiring from the highway. As we wandered up to the door of the building, I felt like I was standing beside something on the scale of a skyscraper, although I would learn later that this telescope is a borderline midget by today's standards.

What do people picture when they hear the word *telescope*? For most, the word calls to mind one of the backyard models mounted on a tripod or maybe the old-fashioned kind that a pirate held to his eye or Galileo balanced on his balcony. Maybe the picture extends to a dome with a literally telescoping tube poking out of it.

People don't generally picture the ten-meter-mirror behemoths that are the twin Keck telescopes on the summit of Mauna Kea in Hawaii, or the Arecibo radio telescope, a gargantuan metal dish nestled in the curves of a Puerto Rican hillside. It's pretty hard to look from a diminutive backyard telescope to the sprawl of giant dish-shaped antennae that make up the Very Large Array radio observatory in New Mexico and see how they can possibly be traced back to the same basic design.

In reality, they can. Most modern earthbound telescopes are built, quite simply, to observe light, and they do so using a series of mirrors. A large, curved main mirror—the primary mirror—collects the light coming from wherever the telescope is pointed and reflects it. The light then bounces up to somewhere else, either a camera or another mirror that continues the reflection sequence, until it reaches some of the world's best scientific instruments, tailor-made to capture the dim light of the stars.

The telescopes themselves are on giant mobile mounts, with motors and gears turning them to keep them pointed toward a single object in the night sky as our planet slowly rotates. Optical telescopes—telescopes designed to observe the same light as our eyes—are housed in domes that block outside light and keep things dark. The tops of the domes are built to turn; the telescope will look at the sky through a wide slit in the dome that turns along with the telescope, keeping the narrow window of sky aligned with wherever the telescope is pointed.

When Phil and I stepped inside the four-meter telescope's dome, the building was cavernously quiet and surprisingly dim after the blinding desert sunlight. There were no lights on, but enormous vents on the side of the dome were open, letting in cracks of light and a breeze to keep the dome interior cool. If the vents were sealed, the whole structure would bake and heat up in the afternoon sun and then take hours to cool off after dark, sending invisible waves of heat skyward and stirring up the air above the telescope like the ripply view above pavement on a hot summer day, interfering with the quality of the observations. There was the quiet murmur of humming machinery, the occasional creak or ping of metal, and the distinctive smell of old motor oil and machine grease seemed embedded in the very walls.

The telescope itself towered in the center of the closed dome, mounted on an enormous concrete support structure painted bright blue. Unlike my parents' old backyard Celestron telescope, housed inside an orange tube, this one—like most modern telescopes—was largely open to the air. Its most important element, the eponymous four-meter-diameter primary mirror, sat at the base of a large white mount, pointed straight up at the smaller secondary mirror held in place by a tall skeletal metal frame. Impressive as it was, the whole setup looked almost small compared to the mass of infrastructure surrounding it: the stairs and walkways leading to a raised platform, the doors in the walls leading to a catwalk wrapping around the outside of the dome, and the shiny metal panels of the dome itself, including the interior mechanics for opening and closing the dome slit.

It wasn't a big-lensed behemoth waiting to be extended and poked

At Kitt Peak National Observatory, opening the telescope dome during my first professional observing run in 2004. *Credit © Philip Massey.*

out of the dome slit, the way telescopes are so often depicted in cartoons. There was no obvious eyepiece at the back, no chair for an observer to sit in. Instead, the back end of the telescope where you'd normally expect to spot an eyepiece disappeared into a forest of cables and wires and metal boxes containing the digital cameras and other research instruments we were about to use.

There was also nobody running around the dome in a white lab coat, shuffling charts or notebooks. The people we did see were members of the observatory's day crew, outfitted in dungarees and T-shirts and more likely to be carrying toolboxes than clipboards as they performed the daily work of keeping the telescopes running seamlessly. There were no star maps or other bits of paperwork scattered about. The overall impression was less sterile laboratory and more garage or construction site. That afternoon, with the vents and dome flung open to the blue sky, the mood inside the telescope resembled the stage of a theater on the afternoon before a show.

The summit of Kitt Peak National Observatory. *Credit: NOAO/AURA/NSF.*

It wasn't quite empty and not quite quiet. The overall feeling was one of preparation, of waiting. The observatory crew slipped in and out, the daylight filtered in, and I got the distinct sense that the whole place was preparing for night to fall and the show to begin, for the (literal) stars to arrive. Astronomy even borrows some terminology from the stage—our time on a telescope is referred to as a run, as in "I have a three-night observing run coming up next week."

That night, when the telescope was observing, there wouldn't be anybody in the dome. The light from the sky—caught by the primary mirror, redirected by the secondary, and gathered by the instruments—would be translated into digital data and sent immediately to banks of computers in the "warm room" tucked away next door, where the observers (professional astronomers) and telescope operators (people specifically trained to operate the great beast of a machine sitting out in the dome) would be sitting and watching the data come in. The dome itself would be cold, dark, and almost undisturbed, interrupted only by the mechanical rumble of the dome turning and the higher-pitched whizz of the telescope following the sky as it moved from object to object.

As we left the four-meter telescope and walked around the mountain to visit a few other domes, I started really taking in the landscape around me. The whole place was incredibly still and silent: wide expanses of dry desert below us, mountains in the distance fading into a bluish haze, the only bit of movement the occasional turkey vulture soaring by *below* the summit. The telescopes were so monolithic and quiet, scattered among the rocks and trees, that they seemed less like construction and more like an organic part of the mountain. I was buzzing with excitement inside, but it struck me just how intensely motionless everything else around me was. An observatory summit is a place of sleeping giants, of getting ready and waiting as night slowly comes.

o o o

CAMBRIDGE, MASSACHUSETTS
September 2002

Arriving at MIT, I was delighted to be surrounded by the camaraderie of a few thousand other science-loving geeks and immediately declared myself a physics major. There was just one catch: I'd never actually taken a physics class before.

I'd *read* about physics, thanks to Carl Sagan and Planetron, and I had a smattering of knowledge about gravity and how stars worked and even some factoids on relativity, but I certainly couldn't have told you much about the math behind how a spring worked, or derived an equation to describe friction, or explained what exactly electricity and magnetism had to do with each other. Still, a physics major was the first step in becoming an astronomer, so a physics major I would be.

I'd been raised on enough inspiring go-getter spunky underdog movies—
Legally Blonde came out the year before I started college—to make this
seem like a *great* idea. *I'll just buckle down and make it work! Sign me up
for the* advanced *introductory physics class! I'm sure I can ace it with enough
determined facial expressions and a "get psyched" playlist!* I'd missed the fact
that those movies usually compressed the actual hard work into zippy two-
minute training montages with lots of drums in the background. It quickly
became clear that they had glossed over the many, *many* nights spent awake
at two in the morning, sprawled on the floor with a mess of notes in front of
you, vision blurring from fatigue, trying to will your homework into making
sense and begging the one person in the study group who knew what they
were doing to not go to bed yet. Physics, as it turned out, was *hard. Really*
hard.

My only consolation was that it at least appeared to be hard for everyone. I have a lasting memory of sitting in that advanced introductory physics course and watching our professor, Frank Wilczek, during one particular
lecture. He was an excellent teacher and brilliant scientist who would win
a Nobel Prize just two years later for his research on quantum chromo-
dynamics, but he would also occasionally forget just how smart he was
compared to us freshmen. One day, he filled two entire blackboards, top to
bottom, with some monster mathematical proof before turning around to
earnestly caution us that "the simplicity of this is deceptive." *Simplicity?* You
could practically see the *we're screwed* thought bubble appear en masse over
the heads of everyone in the class.

At the same time, I adored MIT. I was quickly making lifelong friends,
all of us bonding over our shared classroom battles and helping each other
through our 2:00 a.m. homework piles as much as we could. I even managed
to squeeze in time to attend my very first parties, explore campus by moon-
light, and start dating a fellow freshman named Dave. An athletic computer
science major from Colorado, Dave was taking chemistry and calculus with
me and seemed to see my enthusiasm for astronomy and programming as

a distinct plus rather than a blemish on my femininity. We clicked early and well, and he helped draw me out of my shell by reminding me that I was now beyond the insular middle and high school world where people shunned me for being smart.

My dorm in particular was the stuff of anarchic counterculture geek dreams. When I turned up as a freshman, the residents were busy constructing a gigantic wooden tower that would ultimately stand more than four stories tall. As it turned out, this violated Cambridge building codes, so after a couple of days of climbing all over the thing and hurling water balloons from the top (it was impressively structurally sound; these were MIT engineers building it, after all), the tower was carefully lowered with much fanfare. In the next four years, I'd help my dormmates build giant catapults, human-sized hamster wheels, and even a roller coaster, all purely for entertainment and made primarily out of two-by-fours and optimism. MIT was my first real indication that the road to brilliance sometimes took a few turns that steered well clear of common sense. It seemed evident pretty early on that we were all getting a college experience unlike almost anywhere else, in spite of—or perhaps *because* of—the bruising we were taking from some of the most challenging science and engineering classes in the world.

Through all this, I remained convinced that despite my battles with coursework, MIT was the place for me. I wanted to be a professional astronomer, despite having only the vaguest sense of what the job actually entailed. I'd worked out early on that it meant being in school for the long haul—most astronomers I'd heard of had PhDs—and that I'd probably be using some very large telescopes at some point, but the details were unclear. I'd seen astronomers on PBS or in movies and imagined people sitting behind some enormous telescope situated inside a dome so they could…do *some*thing. It looked fun, and I'd liked our backyard telescope, so I filed this away as something I'd surely figure out when the time came.

In the fall of my sophomore year, I registered for MIT's observational

astronomy course, taught by Jim Elliot. At the time, he was just Jim—
with his encouragement, I nervously got over calling him Dr. Elliot—so
it took a while for the true value of this experience to sink in. Jim was a
groundbreaker and pioneer in observational astronomy and a legend in
his field. The stories I heard from him sounded like some sort of madcap
adventuring cowboy version of astronomy. He had discovered Uranus's
rings and Pluto's atmosphere from the Kuiper Airborne Observatory, a
telescope that observed from the *open door of an airplane*, and had shepherded a veritable who's who of famous astronomers through this very
observing class. His reputation was wildly at odds with what we students
saw as a brilliant but unassuming and friendly professor in his early six-
ties, calmly coaching us through our beginner-level observations and
teaching us the fundamentals of how telescopes worked. Still, hearing
about his observing adventures was a revelation. I'd assumed astronomers were an indoor sort, hunkering safely down inside a dome or puttering over a computer. That I could be a scientist and an *adventurer* was
a new and compelling idea.

As part of the class, Jim would take us out to Westford, Massachusetts,
and the tiny George R. Wallace Jr. Astrophysical Observatory in the evenings for the laboratory hands-on component of our class. Less than an hour
from Boston, Wallace wasn't exactly a dark-sky site—it really wasn't all that
different from my childhood backyard—but it was a bona fide observatory,
with two decent-sized telescopes sporting twenty-four-inch and sixteen-
inch mirrors in their own domes and a shed housing four fourteen-inch
telescopes equipped with digital detectors. The fourteen-inch telescopes
were what we would use, as a class, for our group observing projects. The
class format roughly matched the format of professional observing: weeks
of advance preparation for a small amount of time at the telescope to get a
few hours' worth of data that we'd then spend more weeks analyzing back
at home. For us, a few evenings spent on observing lab trips would give us
what we needed to complete our class projects for the entire semester. For

professionals, a handful of nights at a telescope could fuel months of work and a research paper or two. We were already learning that contrary to popular belief, astronomers spent less time than might be expected at the telescope and more time delving into the data they had gathered during their brief visits.

This ratio sounded fine to me. Going into Jim's class, I hadn't even been sure I wanted to *be* an observing-focused astronomer. I'd never been a terribly hands-on disassemble-the-radio kind of child; I was much more interested in what the telescopes were pointing at than how to point them. I'd always assumed I would stick with the purely cerebral, theoretical side of astronomy (which I imagined meant pondering the mysteries of black holes while tipped thoughtfully back in an office chair somewhere). In my mind, the truly noble pursuit was actually understanding the fundamental physics of the stars, not mucking around with big clunking machines like a mere (and here you can insert the superior sniff of a nineteen-year-old who thinks she knows everything) *engineer*.

It took just one night of observing for me to get hooked. I loved it. Loved gearing up and heading out into the cold clear autumn nights, loved juggling a log book and old computer and flashlight with frozen fingers, loved climbing a ladder and wrestling one of those fourteen-inch telescopes to point it perfectly toward a star of my choice. I loved the thrill that came when everything was working and I could leap back off the ladder to peer at some brand-new data and my own hastily scribbled notes, all under the dim red lights of the shed. (Many observatories use deep-red lighting at night to help preserve observers' dark-adapted vision.)

I have an abiding memory of standing in the November midnight cold, reveling in every perk of my teenaged metabolism as I downed Reese's peanut butter cups by the handful, and peering through the viewfinder of my telescope at the exact moment that a meteorite went streaking through its field of view from top to bottom. I was pointed at a minuscule area of the sky, and the odds of a meteor passing through that tiny spot at the very moment when I'd pressed my eye to the eyepiece were vanishingly small. I don't remember crying out or saying anything or even moving. I just stood there, perched on a ladder, eye pointed through the telescope, knowing what I'd just seen.

Yes, I thought. *This is a good job.*

o o o

KITT PEAK NATIONAL OBSERVATORY, ARIZONA
May 2004

Phil and I ate dinner in the cafeteria with the collection of other astronomers and telescope operators who would be observing on the mountain that night, just before sunset. We all started by picking up our midnight meals, which we had ordered earlier in the day. According to an observer's schedule, the dinner we were about to eat would be the second of our day, and around midnight or 1:00 a.m., it would be time for the third and final meal, referred to by everyone as "night lunch." It's simple fare—a sandwich, a few cookies, maybe a thermos of cocoa or soup—but it helps keep everyone going through the early-morning exhaustion.

I slid into my seat alongside the other astronomers, and Phil introduced me as a summer student going on my first observing run. This seemed to transmit some sort of invisible bat signal among the rest of the table, and everyone started to chime in at once to welcome me, wish me a good night, and offer a few friendly tips that almost immediately morphed into anecdotes and stories of observers who had come before.

"Everyone gets pretty tired around 3:00 a.m., and you can do silly things. I remember a guy who was observing alone and got himself locked in the bathroom one time. He lost half an hour of telescope time before he could get out! Was that here?"

"Not sure, but I do know someone who was observing at the solar

telescope and decided to stick some paper into the beam. You know how
you can do this with regular telescopes to see a little focused image of whatever you're pointed at? Well, this guy put a sheet of paper into a beam of the
focused light of the sun. Burst right into flame."

"Remember to keep an eye out for scorpions. We had an observer get
stung by one a little while ago! Crawled right up her pant leg while she
was at the telescope. Didn't she have to get helicoptered to Tucson for
treatment?"

I must have visibly blanched at the scorpion story—coming from
Massachusetts, the worst I'd seen was the occasional wasp or roach—
because someone in the group decided to continue in the same vein. "Yeah,
the scorpions can be a problem, but have you ever heard the one about
Steve and the raccoon? One jumped right into his lap while observing at
the hundred-inch telescope." *Hundred-inch?* "They said you could hear his
scream as far away as the sixty-inch!" *Where?*

"Forget the critter stories. Who wants to tell the one about the telescope in Texas that got shot?" *WHAT?*

And so on.

(The scorpion sting victim was real but most certainly didn't need a
helicopter rescue. Someone *did* set a piece of paper on fire in a solar telescope, but not at Kitt Peak. There is a Steve who had a close encounter with
one of the well-fed tame raccoons at Mount Wilson Observatory while
observing at the 100-inch telescope, but he says it only tugged on his pant
leg and swears he didn't scream. However, the observer locking himself in
the bathroom was completely true and later immortalized in the methodol-
ogy section of his own research paper, and there is a telescope in Texas that
has been shot.)

This was my first introduction to the spectacular and sometimes exag-
gerated tales of astronomy. Scorpions aside, I was captivated, equally eager
to stay at dinner all night to hear more and to sprint off to the telescope and
see about acquiring some great stories of my own.

○ ○ ○

CAMBRIDGE, MASSACHUSETTS
January 2004

Back at MIT, now hopelessly addicted to astronomy courtesy of Jim's class, I kept pulling out and dusting off the sense of excitement I'd gotten from observing as I continued to bushwhack my way through progressively thornier physics classes. I got some comfort from the fact that most of my classmates were in the same boat as me—we'd all gone from smoothly acing our high school classes to clawing our way to Cs and Bs—but it was still a slog.

Happily, I had at least aced Jim's observing class, along with a field camp experience the following winter that I'd signed up for the moment Jim mentioned it. During January—MIT's short winter semester—he took a small group of us to visit Lowell Observatory in Flagstaff, Arizona, where we would do supervised research with a local adviser and explore the area. (Jim shepherded all the field camp students through a multiday hike down the Grand Canyon, stargazing with us at a campsite by the Colorado River and cooking us pancakes in the morning.) At Lowell, I worked with a young astronomer named Sally Oey. I was somewhat in awe of her. She'd recently won a prestigious national research prize and a competitive funding grant, and she was also a down-to-earth young woman with short hair and cargo pants like mine, clearly excited about our project studying the hydrogen gas in galaxies that could serve as the first seeds of newborn stars.

Sally was traveling frequently that January (something I would quickly learn was the norm for early-career scientists), and I happily hunkered down in her office with the data and research tasks she had given me. Several weeks later, I emerged excited to present the results of what I'd found. For whatever reason, I loved giving talks like this and was fairly decent at it; I tracked it back to years of being on stage as a violinist and drama club geek. My research presentation apparently made a strong enough impression for Phil Massey, another astronomer at Lowell, to choose me as his student when I applied to return to Lowell for a summer internship.

It was an auspicious start to my research career. My semi-arbitrary red vs. blue decision when choosing a summer project would ultimately kick off fifteen years of research on dying stars and a lifelong friendship with Phil. Unbeknownst to us, hidden in the list of stars we planned to study during the summer were the three biggest stars ever observed in the universe, humungous record-setting red supergiants that, if placed at the center of our solar system, would stretch well past the orbit of Jupiter. The mind-blowing discovery, which wouldn't emerge from the data until I'd been through a two-month crash course in observing, data analysis, and introductory stellar physics, would make international headlines and become part of my very first scientific paper. Buoyed by an exciting research project, I would go on to earn my bachelor's degree in physics at MIT and then my PhD in astronomy at the University of Hawaii (unwittingly following in the footsteps of Heidi Hammel, the excited young Jupiter observer I'd admired on TV back during the comet impact in 1994) before beating the brutal odds of the academic job market several times to work as a researcher at the University of Colorado and then a professor at the University of Washington.

I knew none of this as I boarded the plane to Tucson for that summer of research. I just knew that I was still desperately in love with the universe, eager for another chance to prove I could study it, and excited to finally find out, for a full two months and starting with my first real observing experience at Kitt Peak, what it actually meant to be an astronomer.

o o o

KITT PEAK NATIONAL OBSERVATORY, ARIZONA
May 2004

Dinner on Kitt Peak wrapped up in time for everyone to head outside and watch the sunset together before scattering to the telescope, a time-honored tradition of astronomers everywhere. If asked, we would all supply some good practical scientific reasoning behind the habit—you get a good glimpse of what sort of night it's going to be, a sense of upcoming weather, the sky quality, and so on—but the basic reason remains that it's simply beautiful. Standing on a remote mountain with the earth stretching out into the distance and slowly spinning away from our nearest star, it's a wonderful quiet moment to enjoy the vastness and stillness and colors as the night begins. On any given evening, I can promise you that scattered across the planet are a few small groups of astronomers, standing on dome catwalks or dining hall patios or even just a stretch of hard-packed earth and pausing in their work for a few moments to admire the simple beauty of the sky.

A few astronomers standing near me mentioned that I should keep an eye out for a green flash. The green flash, they explained, is an optical phenomenon visible as the sun sets below a very clear and flat horizon. The atmosphere bends the light from the sun as it passes through—an effect known as refraction, which has become a key element of how telescopes work—and separates it into its different colors. At the extreme angle of a sunset, this refraction directs green light toward observers at the last moments before the sun drops below the horizon, giving the final sliver of sun we can see a bright green tinge. "It's better in Chile," everyone on Kitt Peak with me agreed, "because you're looking over the Pacific Ocean. Here, it's much harder." Still, everyone swore they'd seen at least one green flash over the desert.

There was no green flash that night, but the sunset was an incredible one. Far from the riot of fiery, glowing clouds and dramatic sunbeams and red streaks that I'd seen before during cloudy or smoky evenings, sunset at Kitt Peak was gentler but no less breathtaking. The sky gently faded from a stripe of reddish-orange at the horizon through a neat spectrum that transitioned from orange to whitish blue to a clear, deep navy as our mountain was turned slowly away from the sun. There wasn't a cloud in the sky, not so much as a jet contrail to interrupt the perfectly smooth shift of color as it darkened over our heads, the first planets and stars starting to pop into view. It was a perfect astronomer's sunset, recognized by one of the people in the group.

"It's gonna be a good night."

CHAPTER TWO

PRIME FOCUS

George Wallerstein celebrated his sixtieth anniversary of astronomical observing in January 2016 precisely how you'd expect: at a telescope. At age eighty-six, George was nominally retired but in the uniquely academic sense of the word: he may have sported a professor emeritus designation, but he still came to work at the University of Washington astronomy department almost daily. Exactly six decades prior, he had observed for the first time at Mount Wilson Observatory in California as a graduate student, shivering in a cold, dark dome and fumbling with custom-prepared sheets of glass—photographic plates—as he loaded them into the back of the telescope's camera. In 2016, his observing was done from the warmth and comfort of his office in Seattle, remotely controlling a telescope at Apache Point Observatory in New Mexico over the internet and downloading the digital data as it was taken. George pointed out that night that he had observed for exactly thirty years using glass photographic plates and for thirty years using digital cameras, perfectly spanning the technological evolution of astronomy in the last century.

The fact that the dates of his first and sixtieth anniversary observing runs matched to the day is more surprising than it may seem. Perhaps one of the biggest misconceptions about astronomers is that we spend all our time at telescopes, working at them almost nightly and rendering us nocturnal. This also helps feed our image as the typical stereotype of nerdy scientists: the imagined astronomer emerges from the dark once in a while to grab some food or coffee, peer around this weird world of sunlight, and then vanish back into a control room somewhere to steer a telescope aimlessly around the sky like a cosmic video game, waiting for something to happen.

The reality is something else entirely. Time on a telescope is sparse and precious currency for an astronomer. Sitting billions of miles from everything we study, astronomers are largely unable to take the subjects of our research into a lab to poke and prod at them. For most of us, all we can do is look, and for most of the cosmos, this can only be achieved at the world's best observatories. These facilities are in high demand: as rare as astronomers may be, observatories are even rarer, with fewer than a hundred top-notch research telescopes around the world. Even a single night at one of these telescopes gives us the opportunity to point at a handful of stars or galaxies that we've been hoping to study for months leading up to our time at the telescope. A successful night of observing may make us the first to capture the crumbs of light—photons—from these objects that have traveled across the universe and into the telescope for us to study. With the data in hand, we'll then retreat back to daytime offices and desks and computers to puzzle out the fundamental science behind what we've seen for weeks or even months at a time before hopefully heading to a telescope again to answer our next question.

The astronomer stereotype of nocturnal supernerds who can't possibly grasp a life on Earth beyond hunching over telescopes at night often swings wide of the mark, especially for George. He's certainly a certified science geek of the highest order: he received the American Astronomical Society's prestigious Henry Norris Russell Lectureship in 2002, recognizing his lifetime of esteemed work studying the chemistry of stars. Still, despite his unassuming demeanor and stature (short, slight, with a full beard and permanently smiling eyes), he's one of those larger-than-life characters who tends to show up in explorer tales rather than research labs.

George was born in New York City in 1930, just months after the stock market crash. The son of German immigrants, he earned his bachelor's degree from Brown University and then served as an officer in the U.S. Navy during the Korean War before going to the California Institute of Technology for his PhD in astronomy. More than six decades later, he's still an active researcher, continuing his work on unraveling the elements found in stellar atmospheres and boggling the minds of each new crop of students in the department with his stories. George is a champion boxer, a licensed pilot, an accomplished mountaineer, and an award-winning humanitarian. He received the United Negro College Fund's President's Award in 2004 for personally raising millions of dollars for the organization and has supported the NAACP's Legal Defense and Educational Fund since the early 1960s. He also possesses the lethal combination of a bona fide photographic memory and a wicked sense of humor. In pretty much any research discussion, George can be counted on to chime in and recite, from memory, exact results from hundreds of scientific papers dating back to the 1930s along with a choice tale or two about the researchers who did the work.

In his years of observing, George has experienced seismic shifts in how astronomy is done that have moved hand in hand with the technological and digital revolutions of the past six decades. The way we observe today is vastly different from how things worked half a century ago: the data is stored digitally rather than on fragile glass plates, telescopes can be operated from afar or even robotically rather than by hand in the dome, and thanks to the internet, observing astronomers can download references, email in real time with colleagues, and even while away cloudy evenings on YouTube from the most remote corners of the globe. Still, some similarities remain. There is and always has been a sense of urgency underpinning every moment spent at an observatory with the sky dark and the telescope pointed up, a tension between the unstoppable streams of light arriving at Earth from the far reaches of the universe and the scrambling scientists trying to capture them.

o o o

People who imagine the birth of astronomy as the moment when Galileo turned a little spyglass skyward can easily be forgiven for not recognizing what astronomy is like today. The small extendable lenses held up to the eye by sailors bear almost zero resemblance to most modern-day telescopes, in the same way that the first room-sized computers have evolved almost unrecognizably into today's laptops and smartphones. By the time George Wallerstein had his first night at a telescope in 1956, telescopes had long since evolved from tabletop models to giant behemoths, collecting starlight and directing it to cameras positioned at various points around massive domes that could rotate as the telescope slowly turned with the planet, keeping its wide reflective eye focused on the heavens.

Astronomer and telescope builder George Ellery Hale made a career of beating his own records and successively building the largest telescopes in the world during the first half of the twentieth century. This culminated in the crown jewel of astronomy in the mid-1900s, a monster telescope at Palomar Observatory in southern California with a mirror two hundred inches in diameter. From its completion in 1948 through today, any astronomer speaking to a colleague could merely say, "I observed at the 200-inch last night," and their colleague would know precisely where they had been, because of course, there was only one two-hundred-inch telescope in the world, and it was at Palomar.

The measurement and name, incidentally, are an undersell. It's hard to think of anything measured in inches as being truly gigantic, but the two-hundred-inch mirror is over sixteen feet wide and weighs fourteen and a half tons. It outsizes most cars and could squish them into scrap. Even today, more than seventy years after it was built, the Palomar 200-inch is among the twenty largest optical telescopes in the world.

It's one thing to know intellectually that big telescopes mean better images, but for me, the reality of this didn't truly sink in until I got the chance to look through a world-class telescope with my own eyes.

One of the most common misconceptions about modern-day astronomy is that most astronomers still spend their time staring through telescopes themselves. In reality, the opportunity to *look* through the world's best telescopes—really look, eye pressed to a small eyepiece—is rarer than you might expect. Today, many of the best telescopes in the world don't even have eyepieces, and we rely on cameras and other forms of digital data to record what they're pointing at. That said, the chance does still come along once in a while.

One evening at Las Campanas Observatory in Chile, several colleagues and I were spending the night on the mountain but weren't scheduled to observe. A telescope operator let us know that since the smallest telescope on the mountain also wasn't observing that night, he'd be happy to put an eyepiece on it for stargazing purposes if we were interested. Delighted, the whole group agreed and headed down to the telescope shortly after sunset.

With a one-meter-diameter mirror, this telescope truly was a shrimp by modern-day standards, but it still dwarfed most backyard telescopes and was far larger than any telescope I'd ever looked through with my own eyes. In my backyard growing up, I'd enjoyed the views afforded by our little eight-inch but understood that they would never be as spectacular as the photographs I saw in magazines or on TV. Colorful bubbles of gas were reduced to faint white circles, nebulae turned from chaotic rainbows into little white smudges, and Saturn was remarkable because I could see the clear outline of the rings, not because it was an epic and massive color image. The excitement was less about the beauty of the images and more about where they were coming from, knowing that these faint fuzzy blobs were thousands of light-years away.

Lining up to get my first look through a one-meter telescope with an eyepiece, I wasn't quite sure what to expect, but the reactions from the other professional astronomers ahead of me sounded promising.

"Whoa!"

"*Ooooh!*"

"Look, you can see *colors*! It's so...*red!*"

We didn't sound like staid and serious scientists. We sounded like any other group of excited stargazers. We may have all worked with electronic data in our day-to-day lives, but each of us had become an astronomer because at some point we'd fallen in love with the night sky, and for most of us this had meant exploring it with our eyes first. The novelty of this view through a research-grade telescope was lost on no one.

By the time it was my turn to look, we were pointed at a star called Eta Carinae. It was exactly my kind of star: many tens of times as massive as our sun, mysterious, and seemingly nearing the end of its life. It had erupted back in the early 1800s, for reasons we still didn't fully understand, and hurled off a bunch of its own mass, giving it a distinctively odd appearance: a huge cloud of gas shaped like two bubbles stuck together with a bright star at the center. During the eruption, it had been easily visible with the naked eye, but even then, it had just looked like a tiny pinprick of light.

When I looked through the eyepiece, I squealed with exactly none of the dignity of a professional researcher. I could see the bubbles with my own eyes! I could see that they were ever-so-slightly transparent, almost tangibly wispy as they surrounded the star. The star itself was a bright gleaming red to my eye, a consequence of glowing hydrogen in its outer layers. It hung there motionless amid a black sky and a scattering of even fainter stars as I continued to stare.

At that very moment, there was a research paper sitting in my backpack that I was in the midst of writing, describing a new theory for how stars like Eta Carinae worked. Our theory might even explain its strange shape. I'd been working on it for months and was incredibly excited about the results. I'd seen plenty of pictures of Eta Carinae before. Still, getting to see with my own eyes something that had only ever previously existed to me as digital images on a computer or scribbled equations in a notebook was even more exciting than I could have imagined. I'd had no idea the one-meter telescope would be this powerful.

The group of us jumped from object to object, admiring other stars and clusters and nebulae and trying to commit every new and spectacular view to memory. It seemed clear that even for professional astronomers, stargazing never gets old.

o o o

Peering through an eyepiece may be romantic, but it's not particularly scientific. The images we see must be accurately recorded and preserved in some way, and the methods for doing this have evolved over time.

Before photography became widely practiced, eyeballing and hand sketching really were the best means of gathering astronomical data. Solar astronomy still references some excellent drawings of sunspots by Richard Carrington from 1859, and one of my research students once tracked down the first recorded reference for a stellar eruption etched into a seventeenth-century globe. Still, by the time Hale's telescopes came along in the early 1900s, we had long since moved on from peering through eyepieces and sketching what we saw to the most modern technology that could be had: photographic plates.

Photographic plates represented state-of-the-art imaging technology at most observatories. The plates were glass squares that could be ordered— Kodak was a big supplier—and shipped to the telescope. They came pretreated with special emulsions of silver halide that would react to light: more photons hitting the emulsion meant a darker image, and after being developed, the plates would produce a pristine black-and-white negative image of whatever was being observed, with a pale background sky and dark stars.

The devilish nature of these plates lurked in the details. Kodak could produce several sizes of plates, but upon arrival, the plates would usually still need to be custom cut to the size of the camera being used for the observations. These sizes ranged from massive seventeen-inch-square plates, used at smaller telescopes with wide fields of view, to petite thumbnails of plates used for observations with large telescopes or specialized cameras that peered deep into tiny patches of the sky. Since the plates were sensitive to light, they had to be cut in darkrooms similar to those used by photographers. Astronomers would carefully remove a Kodak plate and then slice it down to size using a diamond-edged plate cutter, largely by feel and all in darkness. Even today, many observers who used plates decades prior can still perfectly mime the motions involved in slicing new plates down to size, and almost all of them close their eyes when doing it.

The cutting didn't always go perfectly; experienced observers could tell by sound whether they were getting the smooth *schwwwsssh* of a clean cut or an awkward cut that risked rough edges or a lopped-off chunk of plate. More than once, an observer would be midslice, hear a *crunch*, and shout "lights please!" to their student or night assistant, illuminating the room to face a broken plate and a bleeding hand.

Lawrence Aller was a brilliant and much-admired astronomer of his day, but fastidiousness was apparently not among his many qualities. One day at lunch, he excitedly brought out a finished plate to show his colleagues, sporting a splendid image of a planetary nebula (a beautiful colored bubble of ionized gas surrounding a sun-like star that has reached the end of its life). As he passed the plate around the table, the other observers dutifully admired the image, but finally one posed the question on everyone's mind: the plate they were holding wasn't the perfect little square of most photographic plates, but oddly shaped and jagged, with a broken corner and a sharp burr marring the sides. What had happened? Aller replied that he could never use the damn plate cutter, so he had resorted to just smashing the Kodak plate on the counter of the darkroom and then feeling around until he hit upon a shard that was the right size.

It was also often advantageous to chemically tweak the plates while in the darkroom to maximize how quickly they would respond to light before loading them into the telescopes. Kodak offered a variety of different emulsions sensitive to specific wavelengths of light—from blue to red and even infrared, past the limits of human vision—but even these weren't sufficiently tailored for astronomers' needs. Depending on the wavelengths the astronomers were interested in, the plates would be baked in ovens, stored in freezers, briefly flashed with light, or soaked in a variety of liquids. Most would fare well after being soaked in distilled water, but observers were forever getting more creative—and less risk averse—in their mission to "speed up" the plates, since faster reactions to light meant shorter exposure times.

Infrared plates were a particular challenge. George Wallerstein recalled soaking infrared plates in ammonia, which supposedly increased their sensitivity by a factor of six as opposed to the factor of three offered by distilled water. The downside to this method, of course, was being locked alone in a darkroom over an ammonia bath. When treating the plates with ammonia, George would take care to find someone outside the darkroom and tell them "if I'm not back in fifteen minutes, please come in and drag me out," planning ahead in case he wound up keeling over due to the fumes.[1] Eventually, the ammonia was dismissed in favor of a more effective chemical treatment: treating plates with pure hydrogen gas. Again, the science gains were fantastic but safety became a concern. Palomar built a special room for this, carefully equipped with sparkless light switches and divested of anything else that could possibly start a fire, but it was nevertheless nicknamed Hindenburg Hallway for the duration of its use. On the lower-tech (and less perilous) end, a senior astronomer at Mount Wilson used to swear that nothing was better for speeding up infrared plates than a good soak in lemon juice.

Finally, after being prepared, the plates had to get loaded into the camera itself. This was once again done in darkness, and a crucial step in the process was placing the plate in the camera facing the right way, with the emulsion pointed toward the sky; if the plate was loaded backward, the observations would be useless. Observers widely learned that the best way to find the emulsified side of a plate was to briefly touch the edge of the plate to their lip or tongue to confirm which side was slightly sticky. Apparently, the silver halide tasted slightly sweet, and some astronomers claimed they could even taste the difference between various Kodak emulsions. The smart observers learned to try and lick the *non*-emulsified side.

Even sliding the plate into the camera was a challenge. Telescope mirrors focus starlight not to a single point but to a plane that can be imagined as a square surface. At some telescopes and instruments, this optical plane was slightly curved rather than flat and was best captured if the plate could be curved as well. This was not a feature offered by Kodak, so many observers wound up in the unenviable position of taking their thin, rigid, precision-cut, specially treated, newly licked photographic plates and carefully bending them to load them into the camera while fervently hoping they wouldn't snap. Most people eventually learned how much force to apply, but almost every observer at one of these telescopes knew the gut-wrenching feeling of their carefully prepared plate snapping in their hands…or worse, hearing an ominous *crack* from the plateholder midway through an observation.

That said, the entire process of preparing and loading the plates was merely the prelude to the observations. Once the plate was in place, the telescope and dome, operated separately, could both be spun into position and pointed at their object of interest. Then and only then would the camera be opened, starting an exposure as light from the sky *finally* started pouring onto the plates.

After an observation was complete, the same plates would then need to be developed: removed from the camera, brought back in the darkroom, and carefully brushed or soaked in chemicals to preserve the images they now held. Observers often developed their plates at the end of an already exhausting night, groping around in the dark and trying not to inhale too many fumes from the developer chemicals; in short, this work happened at a time when handling delicate pieces of glass may not have been the best idea. Many an astronomer managed to shatter a post-exposure plate containing hours' worth of work (and more than one then went on to doggedly develop the shards in the hopes that some data might be salvageable).

Underdeveloping a plate risked a subpar image, but overdeveloping a plate was also enough to destroy the data on it, so the developing itself had to be done in a very precise window of time. This was usually fairly routine as long as the observer paid attention but occasionally got a bit more challenging. Paul Hodge was observing at the Boyden Observatory in South Africa and, on his last night, put his entire night's worth of plates into the developer before stepping briefly out of the room. When he opened the door to head back in and retrieve his plates from the developer bath before they overdeveloped, he happened to look down and spotted a cobra slithering into the room ahead of him. Paul froze briefly. Did he surrender the room to the cobra and overdevelop? Did he flick on the lights, ruining the plates? Or did he follow the cobra in and finish developing the plates in the dark, now with a deadly snake somewhere in the room? He opted for the latter, successfully finished developing the plates, and then flicked on the lights to spot the cobra curled up beside the pipes of the room's sink, just next to where he had been working.

Finally, finished and developed plates would be boxed up and brought back to wherever the observer was based so that they could be carefully analyzed. Once again, this process was easier said than done and not infrequently ended with astronomers wincing as a big box of plates rattled around in the cab of a truck driving down the mountain or flying back in coach from an observing run with the boxed plates snugly belted into a first-class seat.

As someone who grew up fully in the era of digital imaging and data, I remember hearing about photographic plates for the first time and imagining them as fairly primitive things, relics of an obsolete observing method with minimal scientific value. This all changed when a friend took me on a tour of the Carnegie plate lab in Pasadena. The plates were *gorgeous*: swirling spiral galaxies, delicate nebula filaments, and exquisite little snapshots of solar system planets, all crisply preserved on thin sheets of glass and as lovely as any Hubble image save for the fact that they were black-and-white negatives. I knew we'd made substantial improvements in our move to larger telescopes and digital data but couldn't deny that I was holding some spectacular (and very breakable) science in my hands.

o o o

The plates, fiddly as they were, didn't bear the brunt of the observing burden. That job still fell to the human astronomers. Observers couldn't just load in their carefully prepared plates and walk away. The cameras the plates were loaded into needed to be operated, and just as importantly, the telescope itself needed to be guided. With powerful telescopes, the sky is so magnified that even over the course of a few minutes, the turning of the earth becomes evident, and the stars the telescope was originally centered on will start to slowly slide out of view. To stay pointed at the right patch of sky, astronomers must constantly guide the telescope, moving and nudging it to keep their object of interest centered. Between loading and unloading plates, opening and closing the shutter, and guiding the telescope in between, most observers simply had to remain at the focus of the telescope during the night, something easier said than done.

Left unimpeded, photons will hit the curved primary mirror of a telescope and bounce back up at an angle, eventually converging to produce a focused image high above the primary. To capture this image, telescopes have a camera—and an associated small cage, just big enough to hold a person—mounted on the top of the telescope struts or tube at what's known as the "prime focus" of the telescope. An observer who needs to operate this camera must therefore get up to the top of the dome, usually via a ladder or small elevator mounted to the wall, and then make their way to the prime focus and into the cage. The methods for doing this can be as primitive as a reinforced board stretched between two walkways, as was the case for the thirty-six-inch telescope at Lick Observatory in central California. There, an observer would reach the walkway, then straddle the board and scoot out into the center of the dome, a solid thirty feet above the floor, to reach the prime focus cage. (The process quickly earned the nickname "walking the plank.") At another observatory in western Canada, more than one first-time observer made their way to the prime focus in the dark via a catwalk, only to then lay eyes on the rickety thing during the day and flatly refuse to make the trip again.

Once installed in the prime focus cage, suspended high over both the dome floor and the mirror, observers were in place to load and unload plates and steer the telescope—which could sometimes tip at a fairly substantial angle—over the course of the night. For both safety and practicality, astronomers would also work with night assistants: while the astronomer stayed beside the camera, guiding the telescope and swapping out plates, the night assistant would be in charge of aligning the open slit of the dome with wherever the telescope was pointing, overseeing large moves of the telescope (say, when moving from a target in the northern part of the sky to one in the south), and generally keeping an eye on things on the ground.

This made good practical sense. Once observers were in the prime focus cage, they generally had to *stay* up there. It was possible to come down, true, but the process was such an ordeal that many observers preferred to just hunker down and wait out the night. This was easier for some observers than others. Many men got into the habit of bringing bottles with them up to the prime focus cage to help answer nature's call without interrupting their observations, and women working at prime focus had to periodically remind their (typically male) night assistants that this wasn't an option for them and they would occasionally need to be let down for a brief break. In some cases, observers would start the night with a thermos of dry ice that they would pour into the camera partway through the night, keeping it as cold as possible and minimizing any errant signal that might be given off by warming camera components. The observers could then put the empty thermos to a somewhat baser use to keep their bladders at bay. (Doing this in the correct order was, needless to say, crucial but not always achieved by sleep-deprived astronomers.)

The chief nemesis of most observers in the dome wasn't bladder control but cold. Guiding the telescope was a delicate and continuous process, demanding the observer stay motionless for hours on end. Winter nights were indisputably the best for observing from a scientific perspective—the nights were long and dark, the cold air crisp and clear—but the misery of spending ten hours shivering away at the prime focus couldn't be denied. The dome interiors couldn't be heated; if they were, the rising heat from the domes would stir the air above the telescope and wreak havoc on the quality of the data they were able to obtain.

Warming the entire dome was a nonstarter, but warming the astronomers was doable. Several observatories purchased a supply of electrically heated flight suits, many of them surplus equipment from World War II pilots. They were a welcome solution for observers but came with their own logistical challenges, since they needed to be plugged in. The suits were designed for 12 volts of direct current power, the sort of thing you could get from a car battery, while standard wall plugs in the United States put out 120 volts of alternating current. At least one observer accidentally plugged himself in the wall, noticed a funny smell a short while later, and realized he was wrapped in a smoldering flight suit.

Even the flight suits couldn't solve everything. Observers would wear the heaviest gloves they could and still be numb-fingered by the end of the night. Their tears would freeze to the eyepiece as they held their eye to it for long hours trying to guide. Howard Bond remembered a particularly cold winter night at Kitt Peak, suffering through nineteen-degree temperatures and forty-mile-per-hour winds that tore straight into the dome and whistled through the prime focus cage, when the telescope eventually ground to a halt. When he called someone in, they realized the grease on the telescope's gears had congealed in the cold, acquiring the consistency of bubble gum and practically freezing the telescope in place, ending the night's observations. Despite the beautiful clear sky and hours of data still to be had, Howard confesses his first thought was *oh, thank God.*

Barring technical difficulties, observers were tied to the telescope until either the exposure ended or the night did. Photographic plates produced truly exquisite images, but even dousing them in ammonia or hydrogen left them far less sensitive than modern-day instruments. Exposing a plate for long enough to get a good image could sometimes take hours or even days. In the latter case, the observer would load the plate, point at their target,

Astronomer Edwin Hubble at the prime focus of the 200-inch telescope at Palomar in February of 1950. *Credit: J. R. Eyerman/The LIFE Picture Collection via Getty Images.*

center the telescope, open the shutter, diligently track the target for the full course of the night, close the shutter, and then leave for a day's sleep with the plate still firmly lodged in the camera. The following night, they'd return, swing back to the same target, center the telescope once more, open the shutter again to expose the same plate, and keep going.

A plate like this was being observed by an astronomer who we'll call Earl (a pseudonym) one evening. Earl was a quiet man, described by some colleagues as bordering on antisocial: entire preobserving dinners with him at the table would happen in silence, and he rarely spoke to his night assistant beyond the basics of an observing run. One evening, Earl was ensconced at the prime focus of the three-meter telescope at Lick Observatory, patiently (and silently) guiding the telescope through the latest in a multiple-night string of exposures on the same plate. Partway through the evening, Earl's night assistant wandered out into the dome, presumably to see how his taciturn observer was doing. As the night assistant stepped through the door, he caught his coat pocket on the dome's light switch. The lights in the dome blazed on, flooding the telescope with light…and ruining the plate.

This was greeted with an enraged howl from the prime focus cage as Earl decisively broke his silence and began shouting curses and promising death and dismemberment to his night assistant. In a berserk fury and still at the controls of the telescope, he began turning the telescope to swing the prime focus cage over to the elevator mounted on the side of the revolving dome, presumably so he could descend and make good on his threats.

This shook the gawking night assistant into action. As far as he could tell, the homicidal astronomer spinning slowly through the air above him might actually be serious. Fortunately, while the observer controlled the telescope, the night assistant controlled the dome, and as Earl approached the elevator, the dome swung into action, turning away from the approaching prime focus cage as the night assistant spun the dome in an attempt to keep the elevator out of reach. The bizarre slow-motion game of circular chase supposedly lasted for a solid half hour as Earl continued to holler and the night assistant insisted that he wouldn't be letting him down until he was in less of a murdering mood. Other astronomers working at the observatory must have been surprised to look over and suddenly see the largest telescope on the mountain spinning in circles, dome open and lights blazing.

Even when nobody was trying to kill anyone, working in high dark places on limited sleep in the middle of the night could lead to some precarious scenarios. George Preston was observing one night at the Mount Wilson Observatory 100-inch telescope, using a different observing setup known as the Newtonian cage. This setup placed a flat mirror that could be tilted to direct light out the side of the telescope near the prime focus instead of directly above it, aimed into a cage that could be attached to the side of the telescope and could hold whatever camera the observer wanted to use. By changing the tilt of the mirror and the position of the Newtonian cage, the whole setup could be arranged such that the observer could sit or stand on a platform hung high on the wall of the dome and work at the business end of the Newtonian cage, loading plates and peering into an eyepiece so they could guide the telescope during exposures. The platform itself could be raised, lowered, extended, and retracted, keeping the observer at a comfortable position relative to the tilting telescope.

That said, the whole setup worked best when the instrument cage could be placed with the position of the telescope in mind. George had positioned the cage appropriately for most of the stars he'd be studying, but he'd also agreed to add one additional star to his program as a favor to a colleague. This star, as it turned out, would require a several-hour exposure and was in a slightly different part of the sky; in fact, it would be passing almost directly overhead.

George, already an experienced observer at this point, started the exposure and informed his night assistant that he could head to his nearby home for a few hours; the telescope would be staying on one target, the dome wouldn't be moving much, and he could come back when it was time to move to the next star. Alone in the dome, with the plate loaded and the shutter open, George took up the usual rhythm of Newtonian observations: look through the guiding eyepiece, adjust the telescope a bit, stand back and wait, look through the eyepiece and adjust again, and so on. As the telescope slowly swung upward—tilting away from the wall of the dome where the platform was placed—George began raising and extending the platform to stay within easy reach of the eyepiece. As the exposure went on, the reach became progressively less easy. The platform was eventually extended as far as it would go, and George began propping one hand on the telescope structure (his weight wasn't nearly enough to shift the hundred-ton behemoth) and leaning over to get at the eyepiece. A few moves later and he was leaning over, bracing his weight against the telescope, and then shoving back against it to tip himself back onto the platform.

The telescope slowly spun higher and higher as the camera exposure continued and the star moved toward zenith, the point in the sky that's directly overhead, bringing the Newtonian cage with it and tilting even farther away from the observing platform. To reach the eyepiece, George was now tipping forward onto his hands and bracing one foot at the base of the Newtonian cage on a small metal flange that helped attach the cage to the support struts of the telescope. This worked well until George tipped forward for his next glimpse through the eyepiece, looked down, and abruptly realized where he was: namely, doing a split between the Newtonian cage and platform a good forty or fifty feet above the concrete floor of the dome.

He was already gripping the cage and had one foot on it, so instinct kicked in: rather than trying to shove himself backward onto the platform, he stepped forward and put both feet on the flange. This left George alone in the dark dome, clutching the Newtonian cage like a scared koala.

His first thought was *I can't have the night assistant come back and find me like this*, followed closely by the petrifying reality of the drop below him. After a few long, frozen moments clinging to the side of the telescope, he was able to make a flying leap back onto the platform (it was only a couple of feet away, but in fairness, they were pretty crucial feet) and salvage both his neck and his reputation with the night assistant.

The design of many of these telescopes made the presence of humans seem like something of a nuisance. Telescopes may have had meticulously polished mirrors and state-of-the-art imaging equipment, but when it came to the observers, the solutions often seemed to be "here, sit on this crate balanced on a wobbly board" or similar. After painstakingly preparing the plates and babying the precious telescope that would be acquiring their data, astronomers would then install themselves high above the telescope, on cold concrete floors, or on a platform at the Cassegrain focus. The latter is the focal position most familiar to today's backyard telescope users, bouncing light from the primary up to a curved secondary mirror and then back down through a hole in the center of the primary to focus an image near the base of the telescope where an eyepiece or camera can be mounted. Even in this scenario, actually staying on the ground was a bit too much to ask for large telescopes: reaching the Cassegrain focus often involved precarious platforms that could be raised or lowered to keep the observer at the height of the focus as the telescope moved, tipping up and down. The platform at the Mount Wilson 100-inch was nicknamed the "diving board"; it was raised and lowered by a chain drive that could occasionally derail and send the board, with the observer perched atop it, free-falling to the floor, a mishap that was supposedly rare but also happened often enough to earn the nickname "the ride." Erica Ellingson remembered once being supplied with a wheeled office chair at a Cassegrain platform. It made for a comfortable seat that abruptly became less comfortable when, in the course of guiding, the chair scooted closer and closer to the edge before eventually rolling off the side and taking a fifteen-foot plunge. (Fortunately, Erica had the reflexes to leap to her feet when the first wheel went over the edge.)

Even observing flat on the ground wasn't ideal. The cold of the concrete floor could seep into people's bones in wintertime, and freedom of movement also meant that astronomers would occasionally forget themselves and attempt to dash across the dome while still wearing a plugged-in flight suit. Sometimes observers still needed a ladder to reach the height of the eyepiece, and old astronomy stories are chock-full of people tipping over or toppling off ladders. Dick Joyce recalled climbing up a twelve-foot ladder to peer through an eyepiece and carefully gripping the ladder the whole time; touching or leaning on the small telescope he was using would have shifted its position. On one cold, dry evening, he ascended the metal ladder, leaning forward to look through the eyepiece, and abruptly got struck with a literal flash of blinding pain: a one-inch electrical shock had leapt from the (grounded) telescope directly onto his (ungrounded) eyeball. In retrospect, he was mostly surprised that he managed to stay on the ladder while reeling from this assault of electromagnetism.

Looking back on all this, you might imagine that the astronomers of this era would have nothing but miserable memories of this sort of observing. It's true that nobody is clamoring for a return to the days of ice-cold domes, fiddly glass plates, hand-operated guiding, and eyeball shocks. At the same time, nearly everyone who has observed like this will invariably cite the time as one of their favorite observing memories.

Once you got used to the height, the cold, and the bladder control, observing in the dome and with the telescope could be a surprisingly pleasant, even romantic, experience. Observers would play music while they sat for long hours, gazing through an eyepiece to guide and swapping out plates. Elizabeth Griffin described a summer night at Haute-Provence Observatory in the south of France, walking between the fourteen domes of the observatory in the clear midnight air and hearing different music wafting out of each one, accompanied by occasional calls of *"C'est fini! Allez!"* as observers finished their exposures and called updates to their night assistants. The backdrop for all this was the dark cool nights, the quiet hum and shift of moving telescopes, and the starry skies overhead.

o o o

The hard-scrabble work going on inside the domes stood in stark contrast to how many of these observers lived on the mountain when they *weren't* at the telescope.

Observatory mountaintops had to be set up to allow astronomers to get sleep and meals, given that most were hours away from any other facilities. Observers could be staying on the mountain for weeks as part of marathon stretches of observing, and even brief visitors would be working on a night schedule and needed somewhere to rest and recharge between evenings. As a result, observatories had to include dedicated living quarters, usually with bare-bones but comfortable dormitories.

Both Mount Wilson and Palomar had dormitories that quickly earned the nickname "the Monastery." There was a clear reason behind the name: at both mountains, women were officially barred from staying in the dormitories and from working as lead observers, a policy that persisted into the mid-1960s. In reality, of course, the female astronomers of the day were unofficially fighting their way into the telescopes from the beginning. In the late 1940s, Barbara Cherry Schwarzschild would observe alongside her husband, Martin, and took on the bulk of the technical tasks, developing plates and guiding the telescope and even flagrantly breaking observatory safety rules to observe completely alone during midnight lunches at the Monastery since she wasn't permitted to attend. Other eminent astronomers including Margaret Burbidge, Vera Rubin, Ann Boesgaard, and Elizabeth Griffin, were all working at these observatories well before women were "officially" granted telescope time, despite being banned from the standard observer lodging.

Dinner at both Monasteries was a carefully dictated affair. The lead observer at the largest—and therefore most prestigious—telescope on the mountain would be granted the seat of honor at the head of the table, with the observer at the next-smallest telescope sitting next to them and so on around the table. Once everyone had arrived—properly dressed for dinner, in some cases—the observer at the head of the table would ring a small

bell, and the cook would emerge with the first course. Further bells would
herald subsequent courses until the whole group had enjoyed a thoroughly
civilized dining experience on this mountaintop in the middle of nowhere.
(At Mount Wilson, a not-infrequent backdrop to all this would be some
observers' underwear flapping on the porch railing, hung to dry after a cursory round of laundry in the middle of long runs.) Once the whole dinner
ritual of hierarchical seating and bell-summoned courses was complete, the
observers would then disperse to their respective telescopes for another
evening of scrambling into cages, slicing their hands on glass plates, and
shivering for hours on end in old flight suits while peeing into thermoses.

Back then, people would often also take an hour-long break in the
middle of the night to reconvene at the living quarters for night lunch.
These dedicated night lunch hours were a chance for observers to stretch
their legs, compare notes, and take a breather. At Palomar, the legendary
astronomer Maarten Schmidt would use the lunch break to play pool with
his night assistant. At the time, he was spending over twenty nights at the
telescope per year. A younger researcher did the math and groused that this
worked out to more than twenty hours of shooting pool, or three nights'
worth of darkness, that could have been doled out to young up-and-coming
astronomers if Maarten would only stop taking those breaks. Another
astronomer, François Schweizer, later argued that the time spent relaxing
and reflecting rather than scrambling to start the next exposure might well
have been the key to Schmidt's greatest scientific feat, the discovery of qua-
sars (incredibly luminous galaxies releasing huge amounts of energy thanks
to black holes at their centers with more mass than a billion suns). I confess
I agree with the younger astronomer; ruminating on the mysteries of the
universe may be invaluable, but it doesn't need to happen in the middle of a
clear-skied observing night, and it's hard not to wonder what an enthusiastic
new observer could have found with the telescope while it waited for the
pool game to finish.

That said, at other observatories, people developed the practice fairly

early on of bringing night lunch to the telescope and eating while they worked. The idea of pausing for an hour during a perfectly good night when people could be getting data was already deeply at odds with the philosophy of some observers. George Preston's thesis adviser, George Herbig (yes, for those keeping track, this brings our George count to *four* for this chapter: Wallerstein, Hale, Preston, and Herbig) was one of several early proponents of the idea that one should not waste a second of darkness and potential observing time. This also translated to arriving at the observatory with a working knowledge of the telescope and a carefully considered night plan. An idle telescope, after all, meant wasted photons and the lost chance to get a slightly longer look at the universe.

o o o

Observing this way—treating photographic plates, clambering into the telescope, and shivering through long nights to gather photons from far across the universe—may have been adventurous and romantic, but it was also physically taxing and time-consuming. The best observers were experts in the technology of the day but also constantly pursuing opportunities to improve.

One major change came in the 1970s with the advent of charge-coupled devices, or CCDs. These silicon chips were far more sensitive to light than photographic plates and capable of capturing far more detail thanks to their ability to convert the light they received into a digital signal. This also meant a tremendous change in how data was stored. Rather than being confined to a single glass plate, digital data could now simply be saved to a tape or disk or server, copied as necessary, and accessed by astronomers at their leisure once they were at a computer.

Newly installed CCD chips along with other improvements established electronics as an inextricable part of life at telescopes. This in turn meant that astronomers were no longer needed to sit beside the camera during

observations; they could now guide the telescope and take images from elsewhere. "Elsewhere" quickly became the warm room, a small room adjacent to the dome with computers, lights, and, blessedly, heat. As computers slowly but surely took over observatories, astronomers were able to spend more and more time in the warm room and less time darting in and out of the dome. Today, almost nobody ventures into the dome during observations. There might be a cursory check-in from a telescope operator (the modern equivalent of night assistants, operators typically know the ins and outs of a telescope's technology better than the astronomers themselves), but the telescope is largely left solo in the open dome as commands and data zip back and forth from a different room.

Telescopes have also ballooned in size as observing technology has evolved. The Palomar 200-inch held the diameter record among telescopes for nearly thirty years but was surpassed by a (problem-plagued) Russian six-meter telescope in 1975 and then conclusively dethroned in 1993 by the first of two ten-meter telescopes being built at the summit of Mauna Kea in Hawaii. Since then, a smattering of telescopes larger than six meters have sprung up at a select few top-notch astronomical sites across the globe, places like Arizona, Hawaii, and Chile. These new telescopes have spectacular capabilities, seeing dimmer and more distant objects than ever before and opening up new corners of the universe, but the downside is there simply aren't that *many* of them. Time on telescopes has become fiercely competitive, and any observer lucky enough to get time isn't taking a midnight billiards break; they're more likely to be scarfing down a night lunch sandwich in front of the computer while trying to wrest a few extra minutes from the evening. George Herbig's observing philosophy, that not one second of time with a clear sky and an open telescope should be wasted, has become a core practice of the field, imparting a sense of value and urgency to almost every moment spent at a working telescope.

It could be easy to imagine the old guard of astronomy remaining set in their analog ways in the classic trope of older generations bemoaning

technology. However, for the most part, the Georges—Hale, Herbig, Preston, and Wallerstein—and their other science-minded contemporaries have been enthusiastically all in on new technology and perfectly happy to swap plate licking and prime focus cages for computers and warm rooms. Even the skeptics were generally quickly converted when faced with the clear gains behind CCDs, automatic guiding, and the ability to instantly work with their data, for one simple reason nobody could argue with: it's better for the science.

o o o

Few people today mourn the loss of fiddly and hard-to-quantify photographic plates, frigid hours in a cold dome, or men-only observatory dorms. The overarching agreement is that the evolution of observing to its current form has been good for astronomy, even as we continue to watch the field evolve at breakneck speed. Still, there is one particular quirk of prime focus observing that's been sad to lose.

Observers working at the prime focus cage would be, quite literally, positioned *at* the focus of the telescope, the point where the gathered light from the sky would be focused into a perfect reflected image, magnified and pristine and ready to be recorded by whatever detector was available. Today, this detector is usually a CCD, and back in the day, it was a photographic plate, but occasionally, the detector available to catch this image was the human eye.

Abi Saha recalled one evening at the Palomar sixty-inch telescope when he was running slightly late, climbing to the top of the telescope after the sun had set to remove a mechanical cover that protected the telescope's mirror during the day. The dome was already open, and the dark night sky was behind him as the cover came off. Looking down the telescope—directly toward the sixty-inch primary mirror—he suddenly came face-to-face with a swarm of lights hovering directly in front of him. They were

tiny and bright, little pinpoints, and as he stared, he realized the strange
floating lights were actually moving in one giant group, slowly drifting
through his vision.

It took a moment for Abi to realize that he was seeing the stars directly
behind him. Their light was being reflected by the sixty-inch, focusing to a
point just in front of his eyes and slowly drifting as the planet spun. Author
Richard Preston described a similar experience in the book *First Light: The
Search for the Edge of the Universe* when he was brought to the prime focus
of the Palomar 200-inch. As he described it, "it seemed as though if one
reached out a hand, one could catch a fistful of stars."[2]

There may no longer be any need to put human eyes at the prime focus;
the science is better, faster, richer, and more beautiful in its own way when
we put cutting-edge technology there instead. The era of sitting in the prime
focus cage with an entire starfield suspended in front of your face as if by
magic seems to be definitively over. (While researching this book, I tried
to find an observatory that would still allow observers into a prime focus
cage during the night, wanting to try the experience out myself, and came
up empty-handed.) Still, lost era or not, observing with a hand on the telescope and the stars seemingly literally at arms' reach is a wonderful sensation to imagine and a great story to tell.

CHAPTER THREE

HAS ANYBODY SEEN THE CONDORS?

Thwump.

The odd sound was enough to get me to shift my eyes a bit to the left, dragging them away from their previously critical job of alternating between the wind speed status window on one of the telescope's computers and LOLCats. It was 2:00 a.m., I was sitting in the control room of one of the two 6.5-meter Magellan telescopes at Las Campanas Observatory, and as evidenced by both the wind gauges and the number of time-wasting tabs I had open in my browser, we hadn't opened the telescope all night. I'd long since finalized and refinalized my night plan, assiduously prepared backup programs for different cloud-cover scenarios, and reviewed old data, but my brain had turned to mush around midnight, and I'd steadily devolved from "work on writing my thesis" to "stare at the telescope's instrument manual and pretend I'm reading it" to "maybe there's a funny animal GIF on the internet that I haven't seen yet." Such is the life of a clouded-out astronomer.

Except I wasn't clouded out, not exactly. I'd stepped onto the catwalk stretched between the twin Magellan telescopes not long ago, and the sky had looked downright glorious: crystal clear, with an ocean of glittering

pinpricks shining overhead. It was precisely the sort of sky every astronomer dreamed of having when working at a telescope.

I was hoping to observe red supergiants again—a direct continuation
of the work I'd begun with Phil back at Kitt Peak—but this time, the telescope was three times larger and the supergiants were considerably further
away, in another galaxy two million light-years from our own. Having been
born ten million years ago (practically yesterday, at least on an astronomical scale), red supergiants still shared a similar chemistry to the gas around
them in their home galaxies. I hoped that by expanding our previous work
to other galaxies, we could see how differences in chemistry affected these
stars' physical properties and death throes. It was another way to ask the
fundamental question at the heart of my ongoing PhD thesis: how did big
stars die, and what did chemistry have to do with their deaths?

If I worked in a lab—if I could just pull a red supergiant down out of
the sky and tinker with it or assemble one out of spare parts—this would
be the equivalent of running two experiments side by side. One red supergiant would have the same chemistry as the gas in our Milky Way, the other
would mimic a nearby galaxy with a little more hydrogen and helium, and
I'd simply build the stars and then speed up time until they exploded to see
what happened.

Instead, since I was restricted to observing light that had left these stars'
surfaces back during Earth's last Ice Age, I was here, on a mountaintop in
the Andean foothills. I'd applied for and been granted time, made it to the
mountain, assembled a careful list of stars to observe, and the telescope was
ready to roll. Everything was working, I was as awake as one could be after
twenty-six hours of travel and an abrupt switch to a night-shift schedule,
and it was a cold and crisp winter night—August in Chile—with the sky in
absolutely perfect shape to get some excellent astronomical data.

It was also *screamingly* windy.

By observatory policy, Magellan closed the dome at wind speeds
above 35 miles per hour. This protected the mirrors from blowing dust and

other detritus and kept wind gusts from buffeting around inside the open dome. We also had to *stay* below 30 miles per hour for at least a few minutes before deeming it safe to open the dome again to ensure the winds wouldn't abruptly pick up again while the telescope sat exposed. Until the wind died down, the dome would remain firmly shut, the telescope awake and humming but pointing steadfastly at the walls. The operator and I had been staring at the telescope's wind status window on our respective computer screens all night. The wind had been holding steady at 40 miles per hour or worse since sunset, but in the past hour, we'd seen hopeful signs of it dwindling. 36…33…31…29!

The thud that had pulled me out of my stupor was the telescope operator's head decisively hitting the desk in front of him as the telescope's wind gauge spiked to 42 miles per hour yet again. I didn't speak much Spanish, but that one was pretty universal.

I was mentally fried and had no idea where I was in a normal human sleep cycle. I had only been granted two nights on the telescope, and six hours into my second night, we hadn't so much as cracked the dome open. If things didn't improve soon, I'd have flown five thousand miles to sit in a closed telescope and browse the internet. I'd be getting a sparse few hours of sleep at the end of the night and then heading home without observing one single photon from a star.

These stars were meant to be an entire chapter of my thesis that would now be lost, possibly forever, since I'd hoped to finish within the year. Of course, that was assuming I even had enough data to show my committee; I knew people who had literally put their degrees, career plans, and personal lives on hold for entire years because of one badly timed cloud. I was a grad student at the University of Hawaii, six thousand miles away from my home state, my entire family, and my boyfriend, Dave, who was in grad school at MIT. Dave and I had wound up dating through all four years of undergrad and successfully weathered a transpacific long-distance relationship in the years since, but we were both more than ready to share a time zone again.

Thrilled as I was to be studying at one of the best departments in the world for observational astronomy, I'd also been doing everything I could to barrel through my PhD program in four years—most people took six or seven—so that we could finally make it happen. It seemed absurd that a windy night in the Andes could derail so many plans and so much hard work.

"Become an astronomer, they said. It'll be fun, they said," I muttered, checking on my rapidly dwindling target list for the umpteenth time. As the wind gusted so hard the building rattled, I couldn't help but wonder what I'd been thinking when I'd opted for this as a career.

"How the hell did I get here?"

o o o

In a literal sense, I had gotten here—here being this incongruously high-tech control room in the Andean foothills—thanks to a solid day of flying through four airports to get from the United States to La Serena, Chile, followed by a two-hour drive out to the observatory itself. It was typical for a "classical" observing run like mine—where the astronomer actually goes to the observatory and stays awake all night overseeing the observations—to start days before the telescope was ever pointed to the sky. Today's best observatories are, by necessity, in the middle of nowhere, and nowhere can often be something of an ordeal to get to.

Air travel usually gets you to within a few hours of most major observatories. Anyone who frequents the airports of places like Tucson, Arizona, or La Serena, Chile, or the Big Island of Hawaii could, by keeping a careful eye out, spot a steady stream of astronomers coming to and from observatories. Dead giveaways for astronomers on their way to a telescope include clothes and laptop bags branded with NASA, observatory, or conference logos, people carrying incongruously warm clothes for the environment, or anyone who looks suspiciously awake late at night. (On the way back, we're easier to spot, especially in sunny vacation destinations like La Serena or Hawaii. Amid the perky sunburned tourists waiting for their afternoon flights, the astronomers tend to be pale, half-asleep, strung out on weirdly timed doses of caffeine, and slumped over in some stolen patch of shade, looking like displaced mole people.)

From the nearest airport, the observatory is at least a couple hours' drive via car or shuttle, which can sometimes wind up being the most eventful step of the trip.

I could fill an entire separate book with stories of astronomers crashing cars at observatories. My brilliant colleagues, most with multiple degrees and PhD-level grasps of physics or engineering, have racked up endless strings of flat tires, cars jammed into ditches or high-centered on rocks, and even a handful of flipped, rolled, or otherwise totaled vehicles that ended with broken bones and trips to the ER. To be fair, it's not entirely surprising, considering the combination of sleep-deprived drivers, rented or borrowed cars, and the roads we're driving. Most telescopes are reachable only by rough-hewn winding mountain roads, built solely to serve the observatories. Since light pollution must be kept to a minimum, there are never any lamps or streetlights. Near the summits of most observatory mountains, signs even admonish vehicles driving at night to turn off their headlights in an effort to avoid beaming light into open telescope domes, leaving drivers scooching cautiously around hairpin turns and switchbacks in the dark. The roads themselves are also minimal at best; some are paved, but others are gravel, dirt, or even just well-traveled wheel ruts.

Observing also frequently transports people with minimal mountain driving experience onto steep summit roads. Telescopes are usually situated at least a little ways above the dormitory and support buildings, so some observatories will maintain a handful of observatory vehicles for observers to use when traveling door-to-door. Still, there's no controlling for, in mechanic parlance, "a problem between the seat and the steering wheel." Several observatories, particularly in the Southern Hemisphere, had to overhaul their vehicle fleet as visiting drivers, particularly Americans, became less and less familiar with stick shifts and sent parked vehicles rolling into buildings or over ridges after forgetting to engage the parking brake. (This prompted the loss of an infamous fleet of VW Bugs that were kept on-site at Cerro Tololo in Chile.) Brake use itself is also key when driving roads like this. Riding and then frying the brakes on steep descents from the summit has sent plenty of drivers careening into dormitory parking lots (or dormitories) with overheated brakes.

The buildings will also occasionally fight back. One telescope in Arizona features an unusual design for its dome: rather than turning the telescope inside the building and then spinning the upper portion of the dome to match, at this observatory, the *entire building* will actually turn whenever it's time to point at a new object in the sky. Tellingly, there's also a clear white circle painted around the outside of the building and a sign warning visitors not to park any closer. The circle demarks the path traced out by the building's protruding stairs as it spins, and for good reason; apparently, someone once parked too close and later got to file the distinctly odd accident report of "a telescope hit my car."

Plenty of astronomers may be reckless, clueless, inexperienced mountain drivers, or simply surprised by telescopes attacking their vehicles, but I'm quite convinced I managed to find the absolute stupidest way to crash a car while observing.

I managed it on the road (sort of) to the Mauna Kea Observatories in Hawaii during one of my graduate school observing runs. I'd flown over to the Big Island from Oahu, where the astronomy department was based, and picked up a little red rental car and a fellow graduate student, Tiantian, to make the trip up to the mountain. Mauna Kea is a mere hour's drive inland from Hilo, on the east coast of Hawaii's Big Island, but it's a hair-raising hour. The winding Saddle Road, crossing over the center of the island, is a roller coaster of small hills and banking turns as it climbs slowly but steadily from sea level to six thousand feet, with a lush tropical landscape slowly morphing into a world of sparse Seussian trees and rumpled swaths of black lava blanketed in low clouds.

At twenty-four, I was intensely aware that I'd paid the extra fees and insurance costs for a young driver. As a result, I was scrupulously careful as I wound through the bizarre landscape, took a right onto the observatory access road, and started climbing into the clouds, keeping an eye out for the cattle that often grazed on the lower slopes of Mauna Kea. An infamous sign on the mountain warned drivers to "Beware of Invisible Cows" for exactly this reason, since they could materialize out of thin air in the fog that tended to blanket the mountain at these altitudes.

At least we didn't actually have to drive to the summit itself; we were headed for the visitors' center and observatory dorms built partway up Mauna Kea at an altitude of nine thousand feet. Tiantian and I were required to stay at the dorms for a night to acclimatize ourselves before continuing upward to work at nearly fourteen thousand feet for several nights. When we did go up to the summit, we'd be riding with the observatory crew. The final stretch of road to the top of Mauna Kea is famously difficult; just above the dormitories, it turns into a narrow, bone-rattling washboard of dust and gravel, and vehicles without four-wheel drive capabilities are officially banned from driving on it. We all knew that every year, a few tourists tried to flaunt this and attempt the drive anyway, and every year, a few tourists got slapped with a bill for the hefty cost of calling in a tow truck from Hilo to drag their rental Ford Fiesta out of a ditch.

As the visitors' center and dormitory building finally appeared through the mist, I was feeling pretty confident. The little red rental car had hung in there (albeit groaning mightily as its Matchbox-car engine battled through the thin air), and Tiantian and I were happily chatting and flicking through music on the MP3 player I'd hooked up to the car's stereo system. We were observing without our adviser on this trip, but so far, all seemed to be going well. It was my fourth trip to Mauna Kea, after all, and after successfully navigating my way through bona fide invisible cow fog, I had decided I was pretty much a seasoned pro.

I then looked down to switch off the MP3 player as I turned into the lot and, while distracted, rocketed the car directly over one of the big triangular curbs lining the parking area, where it landed with a *crunch*.

Well, shit.

Nobody was hurt, though Tiantian was undoubtedly wondering what the hell had gotten into me. Still, after switching off the engine, pulling the emergency brake, and jumping gingerly out of the vehicle, the extent of the problem became clear. I had managed to high-center the car on what was quite literally the only curb for miles around, with one wheel actually hovering in midair. I knew just enough about cars to decide that trying to back it up would be a bad idea.

The sound had attracted the attention of a couple of Mauna Kea park rangers in charge of overseeing daily activities on the upper reaches of the mountain. They took one look at the car and agreed I'd need to make one of those famously expensive tow truck calls to Hilo if I wanted to get the car off the curb without further damage. I groaned, took one last hopeless look at the semi-hovering car, and started dialing. I was mortified. I was going to get in trouble with the rental car company *and* the Hawaii astronomy department that had made the reservation *and* probably the observatory somehow *and* oh god, this was going to be so expensive *and* what if dealing with all this messed with our observing plans *and* while I'd definitely heard stories of people crashing cars before, surely this meant I was *never going to graduate…*

To my amazement, I'd barely connected the call when a tow truck materialized out of the fog and veered into the parking lot after spotting the akimbo red rental car. In a lucky break, the truck driver had already been on the mountain to deal with a dumb tourist who had taken their rental illegally up the summit road; he'd just gotten a bonus dumb astronomer out of the deal. Tiantian took one look at the spontaneously appearing tow truck driver and his offer to heft our car off the curb and dubbed him "Superman," reasoning that only Superman could appear out of nowhere and then lift a car.

Not daring to ask how much it might cost (it wasn't like I had the option to shop around), I let Superman get to work as the fog continued to thicken and the sun went down. Eventually, the little red car was safely back on all fours, and I, Tiantian, the rangers, and Superman all peered curiously underneath to assess the damage. Incredibly, the oil pan and engine were intact, and there was nothing but a small scrape along the frame and the bottom of the bumper. Still, I worried about what to tell the rental car company and asked the group at large if I needed to fill out an incident report or similar.

I'd clearly forgotten I was in Hawaii, land of the laid-back. Superman, to be fair, seriously considered my question for a moment and then slowly replied, "Well…all that really happened is the front got a little scraped, yeah? If they ask, I'd just…say you got a little too close to the curb, yeah?"

I looked back at the rangers, who were nodding sagely at his assessment. Well, sure, "too close" wasn't *wrong*, exactly. "On" was certainly too close. I started nodding along with them as Superman continued. "Really, every rental car has a scratched bumper, so they probably won't say anything. If they ask, you just tell them it was caused by a curb, yeah?" Yeah.

Superman kindly charged me a mere $65 for the tow since he'd already been on the mountain and vanished back into the fog. I executed the most careful parking job of my life to get the little red car into a nearby spot with no further curb encounters, and Tiantian and I headed indoors for dinner. I got a few questions over the next couple of nights from fellow astronomers who had spotted our predicament (mostly variations on "How?!") and knew the story would probably follow me back to the department, but in the balance of things, the most dinged-up thing in the story ended up being my pride. At the same time, I was mildly amused to join the long and illustrious line of astronomers who had gotten into some kind of accident en route to an observatory summit and to manage it with a story that had luckily evolved from "I wrecked a car on the only curb on Mauna Kea and wound up spending a month of my grad student salary to fix the problem" to "There was a curb. Then there was a scrape."

o o o

My trip up to Las Campanas Observatory for my windy observing night had been considerably less eventful. I'd been picked up at La Serena airport by the observatory shuttle and ridden the two hours through the Chilean desert, gazing at the Pacific Ocean out one window and sparse stretches of desert spotted with small cacti out the other. The Las Campanas summit was also fairly compact as far as mountaintops went, with the telescopes a mere fifteen-minute walk up the mountain from the dorms. Since I hadn't yet learned how to drive a stick shift car, I was happy enough to trek back and forth on foot between the telescope and the comfortable but simple dormitories.

After my arrival, I'd had one evening to kick myself onto an observer's sleep schedule. This involved waking up around noon (always the hardest part for me as an incurable morning person), eating a very lunch-like breakfast in the dining room of the main observatory lodge, and placing an order for my night lunch, a sack lunch that I'd pick up at dinner, bring to the telescope, and eat in the control room. During the day, there's often very little for the astronomers on the mountain to do apart from preparing plans for the night and getting rest. Observatories during the day are the domain of the technicians and engineers on the day crew, busily buzzing around the summits during the late morning and early afternoon to check on the telescopes and instruments and make any changes that are necessary from night to night like swapping cameras, cooling detectors, or repairing ornery equipment.

Although Las Campanas was home to only four telescopes—the twin Magellan 6.5-meter telescopes, a 2.5-meter, and a 1-meter—the mountaintop was populated with enough astronomers, telescope operators, and observatory staff to make dinnertime a chatty group affair, even though the Monastery era of hierarchical assigned seating and course bells was long a thing of the past. The food, like most observatory dining, was generally basic but solid fare: meat, grains or potatoes, soup, and the occasional pasta or plate of vegetables for folks with special dietary needs. The exception at Las Campanas came on the famous empanada days, Sundays when the kitchen would churn out delicious heaps of golden-brown empanadas and even toss a few in with your night lunch if you knew to ask.

The dining room offered a funnily incongruous view: salt shakers and ketchup and napkins lined up along dark wood tables and, visible through heavy pleated window curtains, wide brown expanses of empty desert foothills stretching out beneath us in every direction. Just out the north window and up the mountain, the Magellan telescopes were visible, hexagonal metallic domes glinting in the setting sun. To the east, the foothills got higher, and a few corners of snow-capped mountains were visible in the distance. To the west, the foothills were lower and eventually gave way to the distant Pacific Ocean. In the distance to the south amid empty lumps of foothill was La Silla Observatory, an outpost of the European Southern Observatories research organization, with its thirteen white telescope domes spread out like a little string of pearls along a ridge. Chile is unarguably the telescope capital of the world; there are so many observatories scattered throughout the western foothills of the Andes that many can be spotted from another observatory's summit.

The other astronomers and I around the table were happy to be eating a hot meal and swapping stories about our science, colleagues or universities we had in common, and the occasional observing adventure. Still, as the sun dropped and the sky darkened, there was a noticeable undercurrent of excitement building around the table, and in ones or twos, everyone at the table gathered up backpacks and night lunches and started the trip to their respective telescopes for the night. The du Pont and Swope telescopes were farther down the mountain, but the observers on the two Magellan telescopes typically shared the catwalk between the two domes to watch the sun sink into the Pacific over the purple-red foothills while the telescopes churned and hummed behind us. There was usually a pleasant moment of stillness during the sunset—from such a remote outpost, the entire desert seemed motionless—but there was also a slight buzz under it all. For everyone working at the telescopes, the day started at this hour, and the darker the sky got, the more it was impressed upon all of us that we had a busy night ahead.

As long as the weather cooperated.

o o o

If you're an observing astronomer, you live and die by the weather forecast. Thanks to the competitive and jam-packed telescope schedules, you don't have the option of waiting for another night if you get unlucky with the weather; you get the exact night you're assigned and that's it. There's no padding a telescope application under the assumption of bad weather either, no "Well, give me three nights because I really only need one but the weather at this time of year always sucks"; by long-standing agreement, proposals must be submitted assuming clear nights and good conditions. You also don't necessarily have much choice as to when you observe. Different parts of the sky are visible at different times of year, so the targets you're studying dictate when you'll be at the telescope. If your objects of interest are up in the summer, you're stuck with short nights, potential monsoons, and possibly inferior image quality (thanks to ripples of heat shimmering up off the summer-warmed ground as the night gets started), while if you have a winter target, you get nice long, cold nights but the increased risk of a blizzard or ice storm.

The phase of the moon also matters. When the full moon is up, it's both beautiful and *bright*, drowning the night sky in pale-blue light reflected off its surface from our sun. If you want to observe a dim blue object, the full moon becomes your nemesis, outshining the very things you want to study. On the flip side, if you're studying targets that are comparatively bright or fairly red, the moon isn't as much of a concern, and in the infrared, the

moon doesn't emit much light at all. This further splits observing time into bright, gray, and dark nights. Bright time, when the moon is full and brilliant, is almost always doled out to people observing infrared light or especially bright targets, while dark time is reserved for programs studying dim objects or blue light that especially need a moonless sky.

Every observer also has their own personal schedules that limit their observing time and need to be taken into account: teaching responsibilities, conference travel, a holiday or family commitment. Scheduling around all these requirements is a tremendous undertaking for the folks running observatories, and trying to predict the weather on top of everything else is downright impossible. In the end, all you can do is apply, receive an assigned night if you're lucky, and head to the telescope hoping furiously that you get a good night.

A good night also means more than just clear skies. The conditions need to be good enough for the telescope to safely open. Wind is just one problem, posing the challenge of blowing dust or sand or snow or other detritus into the dome. Damp air or fog is just as deadly; a telescope won't dare open if there's any risk of water condensing on the mirror. Thick, low clouds can completely obscure our view of the night sky, but even patchy clouds or high cirrus clouds can cause problems: stars will wink in and out of view or be obscured behind just enough cloud to appear dimmer than they really are.

The magic word for a good night is *photometric*: no visible clouds, only tiny variations in how transparent the atmosphere is to starlight, and good enough for observers to get images that can be considered accurate representations of what's actually there in the sky, behind the atmosphere. Good observing nights should also have good *seeing*, a term referring to how sharp our images are. The little turbulent ripples and eddies of our atmosphere that cause stars to twinkle so prettily translate into a significant problem for observers since they make stars appear blurred in images. Seeing is a size measurement of how blurry the stars are from moment to moment and one

of the most important numbers for observing astronomers. Everyone roots for a low number, with a star as unperturbed as possible by the air.

So after being assigned a night—if you're lucky—based on moon phase, existing travel plans, and the season when your targets are observable, you're left hoping that on your specific evening, there'll be no wind, no rain, no fog, no low clouds, no high clouds, and if the planet's atmosphere could stay as still as possible over your mountaintop, that'd be great too.

With all these requirements, you'd think most astronomers would become amateur meteorology enthusiasts. Some people do turn into forecasting connoisseurs, but many astronomers, including myself, tend to take a more fatalistic approach, semi-ignoring advance information about the forecast simply because there's not much you can *do* about it. People will sometimes assemble cloudy-night plans—backup objects, often for a different research project entirely, bright enough to be observed effectively even in suboptimal conditions—but as often as not, you're just stuck hoping it's clear enough to get your data. The forecasting tricks astronomers do use also aren't always particularly scientific. I had a couple of observers at Cerro Tololo in Chile swear by the presence or absence of Andean condors, immense birds that can regularly be spotted soaring practically at eye level around the summits of most Chilean observatories. According to these observers, spotting condors in the afternoon meant you'd get bad seeing that night; the story came complete with elaborate hand-waving explanations about the thermal currents the condors would be riding, but as far as I know, nobody has ever gathered data on the phenomenon.

Astronomers' helpless dependence on weather has spawned a wacky cocktail of tricks and superstitions, amusingly incongruous among such scientifically minded folks. One colleague has lucky observing socks she dons for every run; another swears eating a banana at roughly the same time every afternoon staves off clouds. People have lucky cookies, lucky snacks, even lucky tables in the dining room they'll sit at before runs. I've developed the strict habit of refusing to check the weather until the day of the

run itself. I tell myself that this forces me to always plan for a clear and pro-ductive evening, but deep down, it's just as much about not jinxing the night as anything else. Some astronomers also seem to have famously bad luck on observing runs. In a few cases, it's gotten to the point where colleagues on the mountain will groan if they see one of their supposedly cursed colleagues on the schedule, convinced their mere presence will summon clouds or rain or high winds and extend their bad luck to every telescope unlucky enough to be nearby.

With only a night or two on the telescope, you also can't just bail if it's bad at the start of the evening. The first half of the night may be cloudy, but those clouds could clear away at midnight and reveal a pristine sky. Following the philosophy that it's criminal to waste so much as a minute of good observing time, this sometimes leaves astronomers camped out in closed domes for hours, chowing down on lucky pretzels and periodically sticking their heads out the door to see if things look better. A scourge of observers is the "sucker hole," a brief patch of clear sky between the clouds that's just enough to excite an observer into opening the telescope. The problem here is that opening a telescope is a bit more complex than popping the lens cap off a camera. By the time you've run indoors, opened the dome, started and prepared and focused the telescope, and swung into position, the hole in the clouds will often have disappeared, leaving you back at square one. Plenty of cloudy or rainy nights consist of astronomers ensconced in domes all over the mountain, patiently waiting it out in the knowledge that even an hour of time with the telescope open and "on sky" can turn the night from an abject failure into a jackpot of data.

o o o

On a clear night, once the sun has been properly seen down, my colleagues and I will all head back into our respective telescopes and get ourselves set-tled for the evening in the warm control room, where a bank of computers

Cartoon by Herman Olivares, depicting a windy night at Las Campanas Observatory in Chile. The speech bubble text reads: "Hi! I'm the Canadian telescope observer, can you tell me the wind speed please?" *Credit © Herman Olivares.*

is set up to open and rotate the dome, turn the telescope, adjust and focus the mirrors, and control the settings and shutters of the cameras that are capturing the data. The data itself is collected digitally thanks to CCDs and is gratifyingly available instantly, popping up on control room computers and getting saved on hard drives as soon as it's taken.

Control of the telescope is almost always in the hands of the telescope operator, who works in tandem with the astronomer for the entire night. Depending on the observatory and the nature of the specific job, telescope operators may have degrees in astronomy, engineering, or both, and all have been trained extensively on how to keep the telescope functional and running. Some of the operators at Las Campanas have worked there for decades. One, Herman Olivares, also works as a professional cartoonist; his art has been published in the national newspaper and adorns the walls of the dining room.

The astronomer may be the person choosing where to point the telescope and working with the data after it arrives, but the telescope operator is the one tasked with making it all happen. Operators are in charge of the welfare of the dome, the telescope, the mirror, and the instruments, and are the ones confirming everything is in good shape, ready to open (or not), and prepared for the night's observations. Astronomers may study the fundamentals of telescopes in school, read up on a particular facility ahead of time, and take on tasks like tweaking the configuration of an instrument or starting and stopping exposures, but the operators are the technical experts and the ones in charge of actually running the telescope from moment to moment.

With the technical challenges in the operator's capable hands, astronomers are left to direct the observing plans for the night, and those plans are often incredibly detailed. I'd been asked by family and friends after several observing runs whether I'd "discovered anything last night!" and the general impression seemed to be that astronomers would periodically deploy to telescopes to serve as lookouts of sorts, waiting with a giant spyglass ready to point skyward at the first sign of something new and exciting like a dramatic stellar explosion or newly arrived comet. While surprises do sometimes happen, a much more typical observing night is one where an astronomer arrives with a carefully designed plan laying out exactly how they're going to spend the next eight or so hours, following a list of targets sorted by importance or brightness or when they're highest in the sky and a step-by-step procedure for when and how to observe each one.

Pointing the telescope to a particular target typically involves selecting it from a list of celestial coordinates the astronomer has provided ahead of time. The operator will then command the dome to turn and the telescope to slew, maneuvering the two giant beasts until they're pointed to the right place in the sky according to the observatory computers.

This method—entering coordinates and then letting the telescope find them—has led to what may be one of the most disappointing disconnects between what astronomers actually know and what our friends think we know: we are *absolutely terrible* at finding things in the night sky on our own.

It's true that some of us are better than others (people who teach introductory astronomy lab classes or who were dedicated stargazers as kids usually do okay), but usually, the best most of us can manage to do is pick out the better-known constellations, remember the differences between summer and winter constellations, and hazard a decent guess at which planets might be visible in the sky. Unfortunately, there seems to be a widespread assumption that "astronomer" means "possesses an encyclopedic knowledge of everything we see in the sky." I've disappointed plenty of people who have asked for the name of a random star only to be met with an "um…" or friends who have asked "Hey, what planet is that?" and gotten back "Er…I dunno…Jupiter, probably?" In astronomers' defense, telescope computers are literally light-years better than we are, combining orbital dynamics and lengthy equations to pinpoint exact sky positions with a precision far exceeding anything we can distinguish with the naked eye. Still, it comes as a surprise to most people that many astronomers can't really find that much in the naked-eye sky.

The "pointing" of a telescope—how good it is at landing on a specific spot in the sky—is usually very nearly perfect, but it still requires a little fine-tuning to get the telescope exactly where you want it before beginning to record data. The telescope will typically be pointed at the right patch of sky and then begin sending back images through the guide camera, a small camera attached to the telescope that takes quick repeated snapshots to give the operator and astronomer a look at where the telescope is aiming. The astronomer is usually ready with sky charts—printed out or saved on laptops from exhaustive online digital archives—to compare what they *expect* to see to what they *actually* see. This tends to turn into pattern matching and a lot of curious head tilting. More than once, I've found myself tipping a printout or laptop at some wacky angle to match the orientation of the telescope on the sky, trying to decide if I'm seeing *this particular* triangle of stars that means my target is nearby. Dim targets in particular are often invisible in the short, quick exposures taken by the guide camera and only appear in the real data after much longer exposures. This double-checking and fine-tuning of where the telescope is pointed may mean an extra few minutes of delay, but it's vastly preferable to accidentally pointing the telescope at the wrong spot in the sky for two hours. When it's already practically criminal to waste time on a telescope, the only mistake worse than not observing is observing the *wrong thing*.

Spending time wisely at a telescope isn't just good science, it's financially prudent. Building a telescope is expensive, with budgets that can reach hundreds of millions of dollars to cover developing remote observing sites, constructing every building on the mountain, and, of course, producing the enormous, finely polished mirrors and state-of-the-art scientific instruments that make up the telescope itself. Yearly operations also require funding and include everything from paying support staff to keeping the lights on. Most observatories are funded through a combination of grants and support from universities, research consortiums, and organizations like NASA and the National Science Foundation. Astronomers who are awarded observing nights typically aren't paying directly for the privilege. Still, the price of a single night in dollars is often pointed to as a tangible illustration of just how valuable telescope time is to the community. Taking both the initial development and ongoing operations costs into account, the world's best telescopes can cost anywhere from $15,000 to $55,000 per night to operate, and their only profits are the scientific strides made possible by their immense observing capabilities.

The ticking clock of an observing run, the very real cost of the night, the preciousness of clear and good skies, and the list of objects to observe all lend the whole process an almost palpable urgency. It's crucial to get to the next target as quickly as possible so we can start collecting data again, but we can't sacrifice a smidge of accuracy. It means an observing astronomer is often a bit on edge while trying to check the telescope's position as fast as humanly possible. Eventually, clock ticking and heart pounding, we convince ourselves that yes, we're probably definitely certainly in the right place in the cosmos, and we can open the camera shutter to begin capturing data. Once that happens, the telescope locks onto its position and carefully tracks our target as the planet spins, leaving us waiting until we've finished the exposure and it's time to begin the process all over again.

o o o

Once an exposure ends, the much-anticipated data from the telescope arrives on the screen, looking…well, terrible. The gorgeous multicolored images we've all admired of galaxies and glittering stars and gas bubbles only look that way after a great deal of effort. Movie science in particular has set us up for disappointment when it comes to how we expect scientific data to look. Real astronomers never get bright red arrows and flashing messages on our computer screens announcing "New Star Detected!" or "We've measured record levels of plutonium!" or "Oh god, we're all gonna die!"—at least not unless an enterprising scientist has written some code and a handy user interface that can automatically accomplish this on the raw data.

Broadly speaking, there are two types of observational data that astronomers can get: imaging or spectroscopy. Imaging is exactly what it sounds like: taking a picture of the night sky. We typically take pictures through filters that tightly control the wavelength of the light we receive, letting only blue, green, or red light pass through the telescope's camera to the detector. This lets us record, extremely precisely, how much light a star is emitting in a specific and narrow wavelength regime. By taking pictures across several different wavelength ranges and then combining them, we can produce lovely color images and use the resulting data to learn a great deal about what we're studying. This sort of data can reveal the shape of a galaxy, how gas is distributed in a nebula, how bright a star is, and exactly where all these objects are in the sky.

Spectroscopy is considerably less photogenic but no less scientifically powerful. In spectroscopic data, light from an object is automatically split and sorted according to its wavelength using a microscopically ruled reflecting surface or prism. (A good everyday example of this is the rainbow effect you can see when looking at light bouncing off the back of a DVD.) The shortest wavelength blue light will be directed to the far left of the CCD, while the longest wavelength light is directed to the right, and intermediate wavelengths are sorted in the middle. Finely splitting the light and then determining how much light we get at each wavelength gives us an object's spectrum, and the instrument that does this is aptly named a spectrograph, since it's essentially photographing a spectrum. A spectrum serves as an excellent means of analyzing something's chemical composition, since light absorbed or emitted by specific molecules or atoms has a very precisely known wavelength. The brightest light from hydrogen will appear yellowish, ionized oxygen appears blue, ionized calcium produces a trio of red lines, and the collective spectrum of any object can be read like a unique fingerprint, giving us a quick glimpse of the physics and chemistry at play in whatever we're observing. We can also use a spectrum to measure how quickly an object is moving through space, how fast it's spinning, or even how far away it is.

Both imaging and spectroscopic data require extensive postprocessing before we can begin to pull any science out of them. Raw data on a CCD chip is obscured by a cacophony of junk data. Electrical noise from the detector, light from the moon and our own atmosphere that the telescope observed along with the target star, and even things as minor as stray heat signatures and the odd over- or undersensitive pixel on the CCD all act as a kind of obscuring fog of bad signal over the real data. Dealing with this is known as *reducing* the data, a term that makes sense when you consider that we're trying to meticulously clear out all the contaminants to get at the scientific data we really want from our object of interest. Doing this is an incredibly delicate art; you don't want to subtract any real signal or leave in any junk, especially because you're trying to preserve the integrity of the data at all times. When beautiful astronomical images are released to the public, they're occasionally met with complaints of "But the data has been manipulated!" In reality, nothing could be further from the truth. It's a similar process to paleontologists hunched over a delicate, just-unearthed dinosaur specimen, using tiny brushes to carefully clear away the dirt and sand to expose the fossilized bone. The underlying science is there, untouched, and looks exactly as it does in reality; we just have to brush away every last grain of electronic sand in order to see it clearly.

This means that instant "Eureka!" moments at telescopes are rare; the data typically needs to be carefully studied before we can say much about it. Still, with practice, most astronomers today are able to at least quickly take care of the basic reduction steps while sitting at the telescope to get an immediate, if rough, look at their data. This is where the nature of digital data becomes valuable: as opposed to the care taken with a one-of-a-kind glass photographic plate or delicate dinosaur bone, astronomers have the luxury of just duplicating a new data file and then doing the computational equivalent of blasting said delicate dinosaur bone with a leaf blower, running the data quickly through some basic processing software that at least clears away enough of the mess to give us a quick peek at what we have. This is invaluable because it gives us the chance to inspect our data as it comes in and make tweaks if necessary, fiddling with things like exposure times and telescope setups to try and get the best observations—and therefore the best science—that we possibly can.

o o o

If you're an astronomer sitting at a properly working telescope on a good observing night, it's easy to settle into a simple and happy rhythm, moving from target to target as you go down your list and doing some leaf blower–style reductions between telescope moves as the data comes in to ensure that all is going well. Astronomer Mike Brown accurately described this sort of work as "the most exciting boring job in the world," and it's wonderfully accurate.[3] A good observing night can be downright dull when everything is going according to plan.

At the same time, it's hard to ever fully forget that you're doing *astronomy*. The files you're calmly copying over to your laptop and puttering with contain zeroes and ones generated by photons that struck a CCD strapped to the back of an enormous telescope humming away a story above you in the middle of a remote desert. The photons themselves may have escaped from the outskirts of a galaxy or the outer layers of a star millions of years ago and have spent that entire time hurtling through space, soaring through intergalactic voids and past distant nebulae and narrowly skirting collisions with objects as large as other stars or as small as interstellar dust motes. At the end of the journey, these photons managed to punch their way through Earth's atmosphere and land, of all the possible places on the planet, on the mirror of your telescope, to be bounced around and funneled into a camera so you could learn a little more about where they came from.

The next time you look at the sky, remember this is the same journey taken by the light from every star you look at that happens to land on your eye.

The journey of the photons you're studying and the immense mysteries of the universe they can illuminate make for an exhilarating and sweeping story. It's a romantic notion to keep in the back of your mind as you settle in for an observing night.

Romance also gets you to about 3:00 a.m. By then, beauty of the universe be damned, most observers can't help but wonder whether the light of the heavens is really quite as beautiful as a pillow.

A half-aware haze sets in during the final hours, especially on the first night of a run (3:00 a.m. on night one is precisely the hour at which the shine wears off for every first-time observing student and they start covetously eyeing the nearest flat surface). The night begins to feel like a long repetitive loop, and the universe starts to feel very big and a bit heavy above your head as you keep grimly cranking through your night plan. You get worse at examining the data and better at reading a book, wandering around the internet, or chatting about anything and everything with your companions. Many observers work in small groups or teams, and 3:00 a.m. telescope conversations are no different from any conversation happening at 3:00 a.m. anywhere else in the world between humans half-drunk on sleep deprivation: a mix of underfiltered and overhonest, meandering through random topics and interspersed with stretches of bleary blinking silence.

The 3:00 a.m. haze in particular is what makes music choice utterly critical to observing runs. Almost any astronomer you ask will tell you that playing the right music is a vitally important ingredient for any observing run, to the point that it acquires an almost talismanic quality. Many observers have music that they only play at the telescope or set up playlists matched to the various steps of the night. Generally, most observers tend toward more energetic music as the night gets later. Someone who might have queued up Bob Dylan at the start of the night will have moved on to AC/DC by the time the early morning hours roll around.

While some people will certainly just tee up Spotify and pop in their headphones, especially if observing alone, the typical tradition is still to curate a playlist and play it in the warm room for everyone to hear. Observing in groups always carries with it the challenge of balancing everyone's tastes, finding the right mix of music everyone might enjoy and music that will "broaden their horizons." The former usually sees observers chiming in together on Gilbert and Sullivan operettas, operators happily tapping their toes along to the astronomers' line-up of Motown or Bruce Springsteen, and groups triumphantly blasting Radiohead or the theme music from

Star Wars as the dome opens. Thanks to the latter, I discovered the Indigo Girls and Utah Phillips on some of my early observing runs, and a friend learned the Louis Armstrong Hot Fives and Hot Sevens practically by heart after an adviser played them on loop during a series of observing runs.

On the flip side, one astronomer started pranking his distinctly biker-dude friend and telescope operator by queueing up Morris Albert's "Feelings" during every run and convincing other observers to do the same on their nights. Astronomer Dara Norman also told me about a run with several other astronomers where they each took turns teeing up music; to shake things up a bit, she slipped in a couple songs by Screamin' Jay Hawkins (an early shock rock pioneer with a macabre voodoo flavor) but was out of the room when the first one came up, returning to some pointedly quizzical looks.

There are also music superstitions: observers will start with a particular song or genre that they think might be lucky for good weather or always end the night with the same wrap-up music. Astronomers span an impressively broad range of music tastes. (It's also worth noting the disproportionate number of astronomers with some degree of musical training, ranging from casual hobbyists to astronomer and rock star Brian May, the guitarist for Queen.) The only general agreement I seemed to get when asking other astronomers about observing music was that you couldn't trust anyone who played gentle classical music all night (too soothing) or anyone who went music-less.

With a combination of coffee, a few last scraps of interesting data, and judiciously applied heavy metal, you can power through to the end of the night, which is usually marked as when the sun is almost up. The trip back to the observatory dorms is always a little surreal. No matter how tired you were during the last couple hours of the night, you're always a little keyed up. You've had a successful night, there's some exciting new data waiting for you, and as you're walking or driving back to the dormitories, it's hard to deny the world around you is starting to wake up. Back at the dorms, you draw the all-important light-blocking curtains (the best are flat-out metal sheets blanketing the whole window to prevent any stray shards of midday sunlight from slicing through) and flop into bed, trying to convince your brain that it's bedtime. Five or six hours later, you're awake again at noon to either catch a shuttle back down the mountain to the airport or start the whole process again.

o o o

Of course, this was what was *supposed* to be happening on my night during that windy trip to Las Campanas, but as I sat in the still-closed dome, I wasn't jumping between targets or downloading and speed crunching data to the dulcet tones of James Taylor. I was glowering at a wind gauge.

I left the observatory while it was still dark out, always a bad sign. At 4:30 a.m., the operator and I had agreed to call it a night. The wind had shown no signs of dying down, and even if it did, we no longer had enough time to open the dome, calibrate the telescope, and get to a target in time to capture any useful data before the sun started coming up. With that, my observing run ended. Thanks to a wind pattern over the Chilean Andes, I had traveled five thousand miles to sit in a closed dome for two nights, staring at an untouched target list and watching my only time on the telescope for an entire year get literally blown away.

I started clomping down the road with one last mutinous glare over my shoulder at the delicate flower of a telescope still sealed up in its dome, cozy as a clam against the wind tugging at my jacket and jeans. I'd come all this way for *nothing*.

It took a few more steps before I started to take in where I was. The moon was down—I'd requested (and lost) two nights of dark time for my observations—and the domes were looming shadows behind me, but the road under my feet still seemed faintly visible. The dining hall and dormitories, far in the distance, were pitch-dark; I'd need to be a lot closer to see the dim red lights placed low along the walking paths to guide people between buildings in the dark without ruining their night vision. Still, I could vaguely see the buildings as well as the curves of the mountains around me and, even fainter still, signs of the higher foothills to the east. On what should have been a lightless night, it took a moment for me to register how I could see anything at all, and when I did, it was enough to stop me in my tracks.

I was seeing by starlight.

The riot of stars overhead on a Southern Hemisphere night is spectacular, particularly for those of us used to the north. Thanks to the tilt of the earth's axis, northern locales see the outskirts of our Milky Way, while those lucky enough to be in the south are looking directly toward the star-stuffed center of the galaxy. The result is a brilliant ribbon of galactic light stretching in an arc across the southern sky. Our galaxy's stars are so densely packed and bright that it's easy to see where they're obscured by interstellar clouds, nebulous blobs that block millions of stars' worth of light and earned the name "dark constellations" from Incan astronomers. They trace out the shape of, among other animals, a toad, a hunting fox, and a mother and baby llama.

When combined with the scientifically mandated utter darkness of an observatory mountaintop, miles from the nearest highway or city, the southern sky goes from beautiful to heart-stopping. Even outside the brilliant stream of the Milky Way, there are so many stars that the sky seems chock-full of them. In areas with more light pollution, we draw lines between visible stars to make constellations and see the areas in between as empty, but as it gets darker, more and more stars start to emerge in these spaces. In a place as dark as Las Campanas, the net effect is a sky so rich with stars, it's nearly three-dimensional. The bright stars are unmissable, pouring photons down to you, and the dim and dimmer and dimmest ones seem almost layered, as though there are even more stars that you can *almost* see lurking just out of view in the darkest patches. The stars also have color in a way that's barely visible elsewhere. The cold white-blues and whispery yellows and pale orange-reds are unmistakable, like looking up into a spilled box of jewels.

I have no idea how long I stood there with my head tipped back, breathless and frozen by the universe. It could have been a minute or an hour. The wind kept buffeting me, but I was anchored in place by the stars above me.

Right. This was why I was here.

CHAPTER FOUR

HOURS LOST: SIX. REASON: VOLCANO.

Being a morning person might be an odd trait for an astronomer, but on the morning of October 15, 2006, I couldn't help it; I was too excited to sleep. I was a newly minted graduate student at the University of Hawaii and gearing up to head to the airport and fly to the Big Island for my very first observing run at Mauna Kea Observatory. I'd be working with my adviser, Ann Boesgaard, to observe some stars in our own Milky Way galaxy. Ann was a renowned astronomer, and under her guidance, we were hoping to determine how much beryllium was tucked into the outer layers of these stars. Beryllium, the fourth element on the periodic table, is something of a puzzle for astronomers who study the chemistry of the universe and the Big Bang. It's hard to make but easy to destroy inside stars and relatively rare as a result. By measuring the small amounts of beryllium remaining in these stars' outer layers, we were hoping we could gain new insights into the stars' life histories and the odd chemical reactions churning away deep in their interiors.

We'd be doing this with extremely high-resolution spectroscopy, sorting out the light from these stars by wavelength at extremely small intervals so that we could identify the precise spot in the spectrum where physics told us beryllium atoms would be absorbing a little extra light. To do this meant using a spectrograph at one of the twin Keck telescopes, the second-largest telescopes in the world with ten-meter mirrors. (Only Spain's Gran Telescopio Canarias, at Roque de los Muchachos Observatory in the Canary Islands, is larger at 10.4 meters.) The Keck telescopes were *legendary*, famously some of the most impressive and scientifically exciting telescopes operating from the ground thanks to their size and their location atop Mauna Kea. The Kecks had taken the first pictures of a planet around another star, set multiple records for the most distant galaxies ever observed, and tracked the motion of stars to prove there was an enormous black hole at the center of the Milky Way. Tonight, I would have the chance to use one of them with Ann, a famous and skilled observer.

The sun was barely up, but I was far too psyched up about beryllium and the Big Island and ten-meter mirrors to go back to sleep. I decided that in lieu of resting up for an all-night run, I could at least pack. By 7:00 a.m., I was pinging around my tiny Honolulu studio apartment with uncontained excitement, tossing and retossing overnight clothes into one of my small camping backpacks with the TV on for background noise.

I was sitting cross-legged on my futon, trying to decide whether I should pack my camera, when the bed started to rattle. I sat up straight and looked around my apartment like a surprised prairie dog as the whole place shuddered for a few long seconds, then stopped.

Huh, I thought. Twenty-two years in Massachusetts had given me exactly zero experience with seismic activity, but after living in Hawaii for two months, I'd been there long enough to at least experience a few small tremors. Still, that one had seemed pretty unambiguous. *I guess that was an earthquake!*

The thought seemed to be the planet's cue for "No, *this* is an earthquake." A Google search later that day for "holy crap I was in an earthquake" (because really, how else is a scientist going to cope?) would inform me that the rattling had been the fast-moving P waves, kindly shaking me to attention before the arrival of the real humdingers, the S waves.

The whole apartment started swaying; the floor jiggled, the ceiling fan rocked back and forth, and the sliding closet doors rattled. The sheer novelty of it had me transfixed for a moment, but when it didn't stop, a quick dose of adrenaline sent me running across the apartment ready to do... something? I lived on the fifth floor, and short of running in circles and hoping my building was structurally sound, there wasn't much I could contribute to the situation. I settled for grabbing my raincoat, keys, cell phone, camping first-aid kit and knife, headlamp, and flip-flops (in retrospect, I have absolutely no idea how I chose so well, since my mental narrative, in its entirety, was *omigod, omigod, what, omigod*), hurling open the door to my apartment, and standing there with a death grip on the doorjamb because I remembered hearing somewhere that you were supposed to stand in doorways during earthquakes. And the building was *still shimmying*. Were earthquakes supposed to last this long, or was the planet broken?

The jostling died down shortly after I'd taken up my post in the doorway. I looked up and down the hall, but nobody else had even stepped out of their apartment. *Well, huh. Must not have been that big a deal.* I headed back into my apartment. *That was intense, but I guess I'm just overreacting.* I tried to calm myself down and return to the task of choosing a pair of lucky observing socks. *I mean, the cable still works, so maybe I'll just...huh, the local news station is out. That's...probably normal.* I unplugged my fully charged laptop and tucked it calmly into my messenger bag as the lights started to darken and the power went out. *Totally normal.* Noises from outside my window told me that people were, in fact, starting to congregate in the street. I'd just added my wallet and folder of observing notes to my bag when the floor starting bouncing again as an aftershock arrived. *Okay.*

I grabbed my bag and my hastily assembled earthquake kit and scurried downstairs, equipped for something halfway between an astronomical observing run and a Boy Scout training video, and met a cluster of building residents flipping through some stations on a clock radio trying to get news. We were mainly getting ukulele music or static, but as we waited, it finally occurred to me: *Is the airport even open? Will I be able to get over to the Big Island? Are we still observing tonight?*

Finally, the radio caught a news station. What we'd felt had been the remnants of a magnitude 6.7 quake. On the Big Island.

The power and cell towers in Honolulu were out, but I eventually managed to connect with some of my fellow astronomy grad students and confirm that yes, observing was indeed cancelled that evening for Ann and me, along with everyone else hoping to get a night of data on Mauna Kea. Since this had been our only assigned night on Keck for the fall, there wasn't much to be done; our spectra and stars and solutions to the mystery of beryllium would have to wait another year. As the day wore on, my friends and I congregated at one of the standard astronomy gathering places, a house near the department that we'd managed to cram eight grad student renters into, and pooled our perishable snacks and battery-powered light sources. We wondered aloud to one another about just how bad the earthquake had been. Had anyone on the Big Island been hurt? Was there any flooding or serious road damage? Could it have damaged the telescopes?

An email was sent to the astronomy department the following day. Luckily, no serious injuries had been reported, but some of the roads to Mauna Kea were impassable, and several telescopes had suffered damage, including both Kecks. Fortunately, the domes and telescope structures had been designed for such things, so none of the damage was catastrophic; the massive glass mirrors and buildings were all still intact. Still, early reports of structural damage meant things would be out of commission, and no observing would be happening for at least the next few days.

The email concluded with "If it's any consolation, the weather here is awful right now."

o o o

Observatories present an interesting conundrum: they're sophisticated and high-tech hubs of scientific activity, some of the largest and best-engineered facilities in the universe. At the same time, they also work best when marooned in the middle of nowhere. By necessity, telescopes and the people who work at them can be subjected to some brutal extremes of the elements. Sarah Tuttle, an astronomer whose research includes building telescope instruments, summed it up well: "We take these high-precision scientific instruments and then just torture them."[4]

The mountaintops where observatories are built are typically remote, exposed, and inaccessible even at the best of times. When combined with the extremes of mountain weather, the environment can sometimes wreak havoc on telescopes. Meyer-Womble Observatory in Colorado, near the summit of Mount Evans in the Rocky Mountains at 14,148 feet, was bombarded by wind gusts as high as 95 miles per hour during the winter of 2011. With the mountain road closed from October through May, astronomers at the University of Denver first noticed something was amiss when one of the observatory's webcams got blown askew by the wind. Further investigation, including photographs by a local mountain climber who was making a winter ascent of Mount Evans in preparation to climb Mount Everest, ultimately revealed that the wind had torn apart the twenty-two-foot dome housing their telescope. (The story ended sadly. After years of effort to replace the dome and battles with contractors, the University of Denver, faced with a lack of sufficient support for restoring the dome, recently decided to demolish and remove the telescope.)

Even milder winds can carry problems. Apache Point Observatory sits high above the plains of southern New Mexico in the Sacramento Mountains but only twenty miles away from White Sands National Monument, an immense span of stark white sand dunes. High winds can send the beautiful white gypsum sand into the telescope domes, effectively sandblasting the painstakingly polished telescope mirrors. Telescopes in the Canary Islands, which lie one hundred miles off the coast of Morocco, encounter a similar problem known as the Calima, a strong wind from the east that carries a thick cloud of dust and sand from the Sahara Desert over the islands.

High on these mountains, severe winter weather and intense blizzards are also a hazard. Sudden storms sweeping in can strand astronomers on the summits if they don't descend in time. Snow and ice on the domes can also seal them shut and become a hazard on their own, since opening icy domes risks dropping broken-off bits of ice down onto the telescope mirrors. Candace Gray, a telescope operator at Apache Point Observatory, recalled spinning the dome of the observatory's 3.5-meter telescope midblizzard to let the driving winds slowly blow off the snow. Taking a more hands-on approach, Ann Boesgaard told me about riding an all-terrain vehicle up to the eighty-eight-inch telescope atop Mauna Kea with some colleagues and then climbing the dome with shovels and ice picks to clear it off.

Being the highest thing around also poses another substantial risk during storms: lightning. Plunking tall buildings on top of mountains is, unsurprisingly, a perfect recipe for lightning strikes. Several colleagues had stories of lightning striking dormitories or support buildings. Even the famed Monastery at Mount Wilson has been hit. Elizabeth Griffin recalled eating dinner there on a stormy evening when a fork of lightning blew apart a nearby fir tree and arced right through the room with such force that it shattered all the windows in the room by sucking them inward.

Dave Silva was observing at the 2.4-meter telescope on Kitt Peak one evening when, in the midst of an electrical storm, the dome itself got struck by lightning. The strike was alarming enough (dome strikes aren't too uncommon, but every astronomer who has experienced one describes it as the loudest sound they've ever heard in their life), and it was followed by the power immediately going out. Dave ran to the dome's electrical closet and flung the door open only to see a huge cloud of smoke billowing out. Convinced the building must be on fire, he took off across the mountain, eventually finding some nighttime crew members, sipping coffee near the summit, who were more than happy to spring into action on an otherwise dull night. Fortunately, the dome hadn't burst into flame, but the smoke wafting out of the electrical closet was all that was left from about eighteen inches of power cable that had been vaporized by the strike.

Rudy Schild, on the other hand, was observing at Mount Hopkins in Arizona one afternoon in 1976. One of only two people on the mountain with a storm approaching, a colleague had called and asked if he could disconnect one building from the main power grid to avoid overloading the entire system in case of a lightning strike. Happy to help, Rudy headed over to the small exposed cage that held the disconnect switch, reassured that the center of the storm was still three miles away but in agreement that lightning was worth planning for. However, the bolt of lightning that hit Mount Hopkins that evening didn't strike a building or anything else on the main power grid; it struck *him*.

An alarm was raised when Rudy's colleague called the other person on the mountain, worried that he had never gotten a call back from Rudy to confirm the disconnect. A search found a pair of glasses and a flashlight lying in the switch cage and Rudy on the ground, boots in the air, about ten feet away.

By sheer luck, an observer who had served as air police in the U.S. Air Force and knew first aid had just arrived on the mountain. That observer found a weak irregular pulse in Rudy's wrist and scrambled to administer oxygen from bottles stored on the mountain. A Forest Service helicopter was summoned, and the pilot ultimately managed to land in the single open patch of land on the mountain, amid the roaring wind and fog of the approaching storm and with only about six feet of clearance for the helicopter's rotors. After being flown to a nearby hospital, Rudy was treated for burns on his legs and feet but returned to work after a few days. In a description of the incident posted on his website, Rudy wrote that the observatory staff were "watching me very carefully for the next few days, but nobody found me any goofier than normal despite my subtle efforts."[5]

Wind, snow, and lightning can all wreak havoc on telescopes, but wildfires are the true terror of many observatories. The dry mountain foot-hills where many telescopes are located are also hotbeds—literally—for forest fires, and while observatories are usually at or near the highest point on a mountain, all but the very highest are still surrounded by enough trees and brush to render them susceptible to fire. Fires in southern California and Arizona have imperiled observatories including Palomar (astronomers on the mountain once got packed into the 200-inch telescope dome and prepared to be evacuated via helicopter if the fire got much closer) and the Vatican Advanced Technology Telescope.

Australian astronomy has been hit particularly hard by fires. Mount Stromlo Observatory, a prolific observatory with historic telescopes dating back to the nineteenth century operating alongside cutting-edge modern facilities, was devastated by the 2003 Canberra brushfires. The observatory lost five telescopes along with workshops, administrative buildings, and homes. (Australian artist Tim Wetherell was later commissioned to create a sculpture, *The Astronomer*, using parts from the telescope ruins, that now stands outside Questacon, the National Science and Technology Centre in Canberra.) In 2013, an enormous bushfire in Warrumbungle National Park caused the evacuation of Siding Spring Observatory, home to more than a dozen telescopes. The fire destroyed buildings on the mountain, but the telescopes fortunately survived and were able to resume operations.

In addition to the stormy and fiery perils of desert mountains, tele-scopes also have a knack for being placed along fault lines—California, Hawaii, Chile—which makes them no strangers to earthquakes.

Every observer who's spent any time in Chile has been through at least one small earthquake. Telescopes actually have something of an interesting quirk when it comes to these tremors: they're pointed so incredibly care-fully and kept so incredibly still that even the tiniest and earliest shake of an earthquake can show up dramatically in the view of the telescope. I remem-ber sitting at a telescope when the operator suddenly exclaimed "Oh! There's about to be an earthquake," a solid second or two before the whole building gave a brief but noticeable rumble. He'd seen the bright star he was using to guide the telescope go skittering right off his computer screen as the incredibly sensitive instrument showed the first signs of being jostled. Since telescopes are specifically designed to be robust against this sort of disruption, the star returned to the dead center of the camera as soon as the shaking stopped, and observations calmly resumed. That said, back in the days of prime focus observing, a few astronomers in California recalled observing mid-earthquake and getting stuck in the prime focus cage for several hours. George Wallerstein explained to me that common practice on these summits was to send the firefighters—who were at least close enough to respond in California—to the biggest telescope first, all in the service of science.

Finally, even volcanoes can get in on the action at some observatories. Telescopes on Mauna Kea have occasionally encountered a phenomenon known as *vog*, a portmanteau of *volcano* and *smog*. Eruptions in Hawaii Volcanoes National Park can sometimes send substantial amounts of sulfur dioxide into the atmosphere, which can mix with condensation to create a mildly acidic fog and lead to lower humidity tolerances for the telescopes. In May 2018, a sizeable eruption from Kīlauea, Hawaii's most active volcano, was captured by webcams on the Mauna Kea summit. Fortunately, the ash from the eruption was blown away from the mountain, and despite vog concerns leading to lowered humidity tolerances at the telescopes, observations were largely able to continue as scheduled.

Since Mauna Kea is situated less than thirty miles from Hawaii Volcanoes National Park, you might think someone there would be able to lay claim to "best volcano observing story." However, this particular distinction most definitely belongs to Doug Geisler.

Doug was a graduate student at the University of Washington, and on May 17, 1980, he spent an exquisite night observing at Manastash Ridge Observatory in central Washington. He was alone on the mountain and taking his very first night of data for his PhD thesis, observing billion-year-old stars in the Milky Way. Early the following morning, he wrapped up observing, closed and covered the telescope as usual, and headed to the nearby dormitory, ready for some solid rest and another fruitful night of science the next day.

A few hours into his "night," around 8:30 a.m., Doug woke up, convinced he'd heard something: a distant low boom or rumble or similar. With nothing apparently amiss, he went back to sleep. He dreamed about the end of the world.

Sometime later, he woke up again and began to prepare for a standard astronomer's "morning": a midday breakfast and a quiet afternoon on a clear-aired sunny mountain. He immediately noticed that something was a bit off: no hint of light was leaking in around the light-blocking curtains in his room. A bit surprised, wondering if he'd epically overslept or if the weather had taken a surprise turn for the worse, he checked his watch— noon—and then decided to take a look outside.

The dormitory door swung open to reveal, at what should have been high noon, pitch-black darkness and a distinct sour brimstone smell in the air. Even armed with a flashlight, he couldn't see more than ten feet in front of him. It was a warm, silent, still day…except the daylight was gone. Doug's first assumption was that there had been a nuclear attack or some similar sort of epic disaster. He was only half-wrong.

That morning, Mount St. Helens, ninety miles west of Manastash Ridge, had erupted, blowing a plume of ash more than fifteen miles high in the most destructive volcanic eruption in U.S. history. The distant sound Doug had heard earlier than morning was likely either the initial twenty-six megaton blast or a deafening secondary explosion produced when superheated material from the volcano instantly vaporized nearby bodies of water into steam. In the hours since the eruption, prevailing winds had carried the bulk of the volcanic plume to the east, right over the observatory and right over Doug.

Like any well-trained observer, Doug kept careful night logs of his observing experiences on the mountain, noting how each night at the tele-scope had gone, any hours lost to weather or technical problems, and details such as temperature, clouds, and sky conditions. Usually, these logs were used by astronomers to remind themselves of the details of the night and by the observatory staff to keep track of any potential problems. Doug's log entry[6] from that day on the mountain has become the stuff of legend:

Hours Lost: 6. Reason: Volcano (good excuse, huh?) Sky Condition: Black + smelly.

I am the last survivor of the war—I remembered the "boom." I rush to the radio—most stations are still playing "cha-cha" music. The end of the world + they're playing "cha-cha" music! Finally KATS in Yakima says Mt. St. Helens blew its wad. I am somewhat relieved. It remains completely dark until ~2, + eventually clears to ~1/2 mile visibility by dusk. I cover the telescopes + instruments. Some of the fine ash is settling thru the slit but I think damage will be minimal. I've heard of the dark run but this is ridiculous.

o o o

Volcanoes and lightning strikes are dramatic reminders to astronomers that we're doing our work on a busy and occasionally volatile planet. It's fairly easy to get buried in our observations and science and forget that the earth is careening through space along with everything else we're observing and that an occasional eruption or electrical storm is just a part of our home's geology and weather. We can also sometimes forget that we're sharing this home with a wide array of other creatures.

Many of our companions on observatory mountains are largely harm-less: the usual cast of squirrels, foxes, raccoons, and small birds that have learned to hang out wherever humans are. Other animals are sheer delights,

and most astronomers will be spotting them for the first time. In southern Arizona, coatimundi, a distant house cat–sized relative of the raccoon
with a ringed tail and mischievous-looking upturned nose, will occasionally
drop by observatories; a few have made it as far as wandering into the dome
and leaving dusty footprints across a mirror. Chilean observatories are regularly visited by guanacos (close relatives of llamas) and owls. A few of the
latter have figured out that observatories' all-sky cameras, small towers with
a fish-eye-lens camera pointing straight up to monitor sky conditions, are
excellent perches for hunting purposes. Astronomers periodically checking
the camera for clouds will occasionally wind up getting an eyeful of fluffy
owl butt or a curious giant-eyed face peering back at them.

Most first-time visitors to Chile are warned by veterans in gleefully
descriptive detail about the tarantulas, residents of the high Andean desert
that find their way into everything with a skill seemingly at odds with their
hefty proportions. Adults are solidly hand-sized, with thick gray-and-black
bodies and the distinctively hairy and high-kneed legs of every arachno-
phobe's worst nightmares. They're also common enough to make pretty
frequent appearances in the cool dark corners and crevices of Chilean
observatories and further complicate matters by being more active at night.
Their ubiquity means that astronomers have unwittingly wrapped their
hands around tarantulas perched on stair railings in the dark, come back to
their control room chairs after bathroom breaks to find one sprawled across
the seat, and tried in vain to fall asleep in the dormitories with a tarantula
perched on the wall above their bed.

I heard so much about tarantulas before my first observing run in Chile
that I became convinced these were scare-the-new-girl-type exaggerations,
only to encounter my first one sitting directly on the door handle of my
dorm room. We were stuck in a brief standoff—I certainly wasn't prepared
to reach my hand toward it—before the tarantula sprang off its perch (and I
sprang backward about eight feet) and scurried off into the desert.

In truth, these tarantulas are actually fairly shy, skittish, and fragile creatures that can be easily hurt if handled improperly. Most regulars at Chilean observatories have achieved, if not a peaceful coexistence, at least an uneasy truce with the eight-legged observatory residents that, after all, are usually content to hang out in a corner of the control room or scurry off in fright if startled. That said, it doesn't make the initial sight of them any less alarming to first-timers.

While they may seem downright cuddly compared to giant tarantulas, miller moths are unarguably the most frustrating and persistent scourge of observatories in the American West, largely due to their ability to find their way into *everything*. One intrepid moth confused the hell out of a group of astronomers who couldn't understand why they were seeing nothing in the telescope's field of view one night despite pointing straight at a bright star; they eventually realized there was a moth perched on the detector directly at the telescope's focal point. Scads of moths can mob the dark, cramped spaces of observatories, choking electronics and clogging motors and drives to the extent that operators sometimes have to crawl deep into the telescope to clean them out. Frustrated astronomers and observatory staff have tried a variety of repelling measures over the years—sound, air guns, flashlights, fluorescent bulbs, lavender oil, and copious cursing— but multiple observatories have settled on what has been nicknamed "the mothinator," a simple but effective combination of a lamp, a fan, and an industrial-sized garbage bucket that can fill to the brim with moth carcasses in a matter of days during peak moth season. Ladybugs can cause similar problems; early every summer, huge swarms of ladybugs traversing the American Southwest will alight on high mountain peaks during mass migrations, turning out in such numbers that entire building walls can appear bright red and slightly teeming.

All of these, though, pale next to the scorpions. At observatories in the American Southwest and Australia, scorpions *do* pose a danger to astronomers. New observers are warned sagely to watch out for the "small brown ones" and told to shake out towels, bang out boots, and give their pillows and bedsheets a once-over before climbing in. Sarah Tuttle was observing at Kitt Peak one evening, noticed a tickling situation on her leg, and quickly realized something was climbing up the inside of her jeans. Thinking quickly, she clamped her hands around her knee and stomped, but the scorpion still managed to sting her before dropping to the carpet. The intensely painful sting prompted a call to the emergency medical technician stationed on the mountain, but fortunately Sarah didn't have an allergic reaction, so her treatment amounted to painkillers, ice, and rest. Still, she and her colleagues spotted several other scorpions during their observing run and thus spent the rest of the run sitting with their feet curled under them or their pants tucked into their socks. The sheer unpleasantness of the story also made it a frequently told tale on the mountain. A few years later, Sarah, back at Kitt Peak for another run, was breathlessly told over dinner about a poor woman who'd had a scorpion *climb up her pant leg* and wasn't that just awful? (The story apparently continued to evolve into the "evacuated to Tucson by helicopter" version that I heard during my first visit to Kitt Peak.)

Scorpions and insect swarms have a freakish ability to sneak in just about anywhere, but larger creatures generally stay away from the noise and smell and activity of observatories…unless they're unwittingly invited in. On one beautiful summer evening at Kitt Peak, someone decided to leave a door open to let in the gentle mountain breeze, a plan for fresh air that backfired rather spectacularly when a skunk wandered in. By the time it was discovered, it had made it too far into the building to be startled away (for multiple reasons). The clever scientists who found the skunk put together a plan, laying a trail of bread crumbs leading down the hall and then a good ways out one of the doors. Skunks, it seems, particularly like bread crumbs. The plan was working great, with the skunk following the bread-crumb trail through the building and getting closer and closer to the open door, until it was stopped in its tracks by coming face-to-face with another skunk that had picked up the trail from the *outside* and dutifully followed it in.

Still, nobody has learned the "don't leave the door open" lesson quite as dramatically as the observers at Apache Point Observatory. The telescopes at Apache Point are controlled from a central building, with separate rooms for different telescopes along with a lounge and kitchen area connected along a long hallway that ends with a door to the outside. On one beautiful morning, the building was all but abandoned, with the door propped open and one last tired but happy telescope operator enjoying the fresh air in the doorless 3.5-meter telescope control room at the end of the hallway. Finishing their work, the operator stood up, stepped around the corner to leave…and came face-to-face with a black bear in the hallway.

Most large animals generally stay on the periphery of observatories, though they'll occasionally make an appearance. Black bears are common in the mountains of the mainland United States but can usually comfortably coexist with observatories; astronomers are simply given flashlights and warned to give bears a respectful berth. Similarly, shining a flashlight out the door in the middle of the night at some Australian observatories will sometimes illuminate rows of kangaroo eyes glinting back at you. In Chile, multiple observers have nearly crashed head-on into wild burros while walking around the observatory in the dark, but these stories generally end with a shrieking astronomer and freaked-out burro scrambling off into the night in opposite directions.

Fortunately, that's how the Apache Point bear encounter ended as well: the operator and bear both simultaneously panicked, and the bear luckily went barreling back into the great outdoors while the observer dove for the nearest control room with a door.

o o o

If astronomers as a community were asked to pick a favorite observatory animal, it would likely be the viscacha. Viscachas are relatives of chinchillas but resemble wise rabbit grandfathers with tall ears, long curled tails, sleepy eyes, and long, drooping whiskers. They frequent many Chilean

observatories, and their steady presence over the years has alerted astronomers to a funny quirk of these little creatures: they seem to love watching
sunsets.

Invariably, when a group of astronomers gathers on a Chilean observatory summit to watch the sun go down, we can spot a viscacha or two somewhere along the hillside. (The populations were supposedly much higher
before the cafeterias started tossing scraps to Andean foxes and upping the
predator population in the area.) They're always there, always sitting stock-
still, and always gazing directly at the sinking sun on the horizon.

It makes for an interesting contrast, sharing our summit and sunset
with these meditative animals. Most astronomers out admiring a sunset
are about to spend their evening, very literally, with their minds thousands
or millions of light-years away. The viscachas, presumably, will spend their
evenings munching on grass or moss. Howard Bond recalled an evening
spent at Cerro Tololo with a viscacha calmly settled just below him, sitting
and watching the sunset together until it got dark. As he described it, "Here
are these two critters in this vast universe watching this celestial show…this
show that may have nothing to do with us…but there it is."[7] From a cosmic
perspective, astronomer and viscacha alike are small living things, perched
on a mountain, watching our home planet turn.

a satellite dish or communication antenna, a parabolic bowl made of white metal mesh. The behemoth was twenty-three stories tall and weighed six hundred tons, but despite all that, it could be steered and focused on the sky with pinpoint precision thanks to operators like Pete, who had all been specially trained to run the telescope. It had observed the birthplaces of newborn stars, was instrumental in the discovery of dark matter, and had conducted an exhaustive survey of the sky to assemble a list of any objects emitting radio light. Its size and bright white color had made it a landmark in the area. Driving north on Route 28, it couldn't be missed, looming over a hill behind a local farm.

Except Pete hadn't seen it. He was scheduled to work the day shift at the 300 Foot on November 16, 1988, and made the drive so often that it took a moment to register what he had—or rather hadn't—seen that morning, but as he kept driving, an oddity stuck persistently in his head. He was sure of it: the telescope hadn't been there. Which seemed impossible.

It wasn't.

In the days before, several operators and mechanics had commented that they'd heard the telescope making noises, occasional snaps or pops or scraping sounds, but none of it struck anyone as particularly out of the ordinary; enormous metal structures like this creaked and pinged all the time. The telescope continued operating as it always had, and on the evening of November 15, it was working fine, capturing new observations for its landmark sky survey. The following day, the telescope was scheduled to undergo some routine maintenance, swapping out receivers to modify the specific wavelengths of radio light that it would record. The process would have involved an operator and mechanic physically climbing the dish itself to make the change.

The story of what happened to Green Bank's 300 Foot telescope that night is laid out in *But It Was Fun: The First Forty Years of Radio Astronomy at Green Bank*, a collection of scientific papers and personal stories chronicling the observatory's first decades of operations. Greg Monk was the operator

CHAPTER FIVE

THE HARM FROM THE BULLETS WAS EXTRAORDINARILY SMALL

Pete Chestnut was halfway to work when he realized the telescope was missing.

Pete was an operator for the 300 Foot radio telescope at Green Bank, nestled in the midst of the United States National Radio Quiet Zone in West Virginia. For radio waves—the longest wavelengths of electromagnetic light, invisible to the human eye—this area was a dark corner of the globe, hemmed in by strict rules on the technologies that could be operated in the region to minimize any production of radio waves. To this day, people within a twenty-mile radius of the observatory can't use cell phones or Wi-Fi networks, and even vehicles in the area are all diesel powered to prevent interference from the ignition sparks generated by gasoline-powered engines.

Green Bank's 300 Foot radio telescope was the single largest telescope in the world when it was built in 1961. Designed to collect, reflect, and focus the much longer wavelengths of radio light rather than optical light, the 300 Foot looked almost nothing like its optical telescope cousins, which were all outfitted with precisely polished primary mirrors and tucked safely inside a protective dome. Instead, the 300 Foot looked more like a giant version of on duty that evening and the only person at the 300 Foot, sitting in the control room of the building directly beneath the enormous telescope. During one of the telescope's standard scans, he stood up from the control panel and started to head down the hall to grab some food from the kitchen.

As he described in *But It Was Fun*, there was a loud crack, then "a low rumble like an overhead jet aircraft and then a crash," followed by something smashing through the ceiling.[8] Ceiling tiles and light fixtures tumbled to the floor, the building lights went out, and he could see a cloud of dust falling down the hall.

Wheeling back to the telescope control panel, Greg punched the emergency stop button, ran out of the building, and jumped into his truck to get help. As the car swung around the parking lot, it illuminated some debris lying on the ground, but Greg's priority was hurrying to the 140 Foot, a smaller radio telescope up the road in the same observatory complex where he knew he would find other employees. As he drove, tinkling glass revealed that the back window of his car had been smashed. Later, someone would spot a large bolt, "painted like the 300 Foot," sitting on the car's back seat.[9]

On the drive back—now joined by George Liptak, the 140 Foot supervisor, and Harold Crist, another operator—the car's headlights revealed the full extent of the damage. The entirety of the telescope—the dish, the support structure, *everything*—had crumpled to the ground and was lying in a chaotic heap. George Liptak described it as "like a rotted mushroom that had collapsed,"[10] while Harold Crist's impression was of "a big steamship collapsed or capsized."[11] Ron Maddalena, a staff astronomer who arrived at the site soon after, commented, "I didn't know steel could bend like that: it looked more like caramel than steel."[12]

The 300 Foot Green Bank Telescope, just before and after its historic collapse.
Credit: Richard Porcas, NRAO/AUI/NSF.

As word spread, how the news was received followed an almost perfectly predictably pattern. Each observatory staff member would respond
with incredulity (*Fell down*? Telescopes don't just *fall down.*), followed by
low-grade alarm as they headed to the site, which morphed into dumbfounded disbelief upon seeing the telescope's remains, and then slowly gave
way to mourning. Anyone who so much as laid eyes on the site could tell the
telescope was utterly unsalvageable. Still, the crew that night worked to preserve the electronics (the roof had gaping holes in it, and there was a chance
of rain) and marveled that nothing more serious had happened. Enormous
beams had punched straight through the building in areas, and it was a relief
nobody had been hurt.

The following morning, after mentally registering the no-longer-
there telescope during his drive, Pete Chestnut arrived at the observatory
at 8:00 a.m. to find a heap of white wreckage. He learned that the telescope he worked at was gone around the same time as the world; word
of the collapse quickly spread and eventually made the national news. In
But It Was Fun, Pete explained that he'd been preparing to buy a house
at the time, and just the day before, he'd gone to the bank to apply for
a loan. He stood with his nonplussed colleagues staring at the wreckage
until 9:00 a.m., when the bank opened, and then drove up the road to an

observatory building with a phone to call them. "Hold the loan," he told them. "I may not have a job."[13]

An intense investigation, auditing construction plans, safety inspections, and telescope data that had recorded right up to and even during the dish's collapse, produced a 112-page report. The catastrophic structural failure was ultimately traced back to an overstressed gusset plate that was a critical element of the telescope's main supporting truss. The investigation concluded the failure was nobody's fault: standard maintenance and operation procedures had happened appropriately, and there had been no previous structural problems that could have hinted at a coming collapse. There was also no particular reason to believe other radio telescopes were at similar risk.

In short—and incredibly frustratingly—the telescope had, it appeared, *just fallen down.*

o o o

The 300 Foot collapse is by far the most spectacular example of a telescope just up and failing on us. While nobody was hurt, it drove home both how incredible it is that we can build and operate these massive and complex pieces of equipment and how easily things can go wrong.

Telescopes represent an exquisite pinnacle of optics and engineering capabilities. They are scientific instruments the size of a house, built to withstand the bombardment of harsh mountaintop elements, but must also be capable of making absurdly precise moves, perfectly tracking the very motion of our planet.

Their mirrors or dishes must be curved to mathematical perfection to properly focus the light they collect, with astonishingly tiny margins of error. This need for extreme precision is a consequence of physics: for a telescope to properly reflect and focus light, its shape must be exact to within 5 percent of the wavelength of the light it's focusing. For optical telescopes, this translates to a precision of about twenty nanometers, thousands of times smaller than the width of a single strand of human hair. On this scale, even a seemingly minuscule error in a mirror's curvature can be enough to wreak havoc on what we hope will be gorgeous data. The Hubble Space Telescope famously launched with an out-of-focus mirror, a problem that ultimately required astronauts to install corrective optics during a servicing mission. The whole debacle was a result of its eight-foot primary mirror being too flat by about 1/10,000th of an inch.

Mirrors are, unsurprisingly, common fail points at telescopes. It's not necessarily for the reason you'd think, namely that they're giant pieces of glass and therefore fragile. They're glass, true, but most telescope mirrors are made of literal tons of thick borosilicate glass, the same robust shatterproof material used in old glass baking dishes. Admittedly, they're not *impossible* to break. Jay Elias is one of the unlucky few who found this out firsthand. Sitting down to breakfast one afternoon at Mount Wilson with two other astrono- mers, George Preston and Anneila Sargent, George asked how everyone's previous night had gone. Jay, who had been observing on the mountain's relatively small twenty-four-inch telescope, responded with "well…a little difficult, maybe."[14] When pressed, he explained that the secondary mirror had fallen out of the telescope, the fault of an incorrectly tightened support plate. Alarmed, his tablemates asked if the mirror had broken.

He thought about it for a moment. "Well, some of it did."[15]

My adventure with the maybe-about-to-drop mirror at the Subaru tele- scope had been another example. I got lucky that evening; the day crew was correct, and the telescope's reported error turned out to be a false alarm, so "turn it off and back on again" was all I needed to do to fix the problem. If I really *had* lost the supports for the four-hundred-pound secondary mirror and sent it on a seventy-foot plunge, it would have certainly broken the glass. It would likely have also put a good dent in the concrete floor.

Still, the biggest challenge with the mirrors is damage to their carefully shaped and shined surfaces, the same concern prompting these telescopes to stay shuttered against snow, rain, blowing sand, and wind. Even with top-notch vigilance to avoid damage, most modern telescope mirrors are periodically removed from their support cradles so they can be stripped, washed, polished, and "realuminized," getting a thin new coating of aluminum or silver to restore their pristine surfaces against normal wear and tear. Some observatories have realuminization equipment on-site, while others have brief closures every few years while they ship their mirrors elsewhere for care and maintenance. Still, whether the mirror is moving ten feet or ten miles, it's always a hair-raising process to heft these giant mirrors out of their telescopes and hope they don't get dropped.

Weather is the biggest culprit when it comes to damaging mirrors, but it's not the only one. A camera mounted at the prime focus of the Subaru telescope sprang a leak one evening and sent the bright orange coolant used in the camera—water and ethylene glycol, the same mixture used in car antifreeze—pouring onto the lower parts of the telescope, including several other cameras and the primary mirror. The spill looked alarming at first, with bright orange liquid spattered across the normally pristine mirror, but luckily, the leak was contained within the dome, and the liquid was noncorrosive. Astronomers at another telescope weren't as lucky when liquid mercury leaked from an inner tube used to balance and support the secondary mirror. A drop of mercury was first spotted on the carpet inside the dome, quickly prompting a thorough cleanup and investigation (complete with a visit from the Occupational Safety and Health Administration). Still, the true surprise came when someone took a good look at the mirror. Mercury and aluminum, it seems, do not get along, and the droplets of mercury falling onto the mirror had stripped away large swaths of the aluminum coating.

o o o

Like any hunk of technology with more than a few moving parts, telescopes can be alarmingly adept at breaking themselves. The rogue gusset plate that took down the 300 Foot may be the most epic example, but it's not alone. Many of the more trivial breakages don't cause any long-term damage, but they can still knock out a night of observing or delay a run. On one of my early runs at Cerro Tololo, the shutter of the camera we were using malfunctioned, and a replacement had to be brought in from La Serena, a two-hour drive away. I remember standing outside the Cerro Tololo dome with my collaborator for hours, admiring the gorgeous sky but also keeping one eye trained on the foothills below us as a lone pair of headlights slowly snaked toward us from La Serena, carrying the shutter.

The moving domes are similarly vulnerable: any errors in their ability to open, close, or turn can stymie observations. As with weather, most observers will opt to sit around and wait it out, hoping that some of their precious telescope time can be salvaged with a quick or creative fix, but sometimes the problems go beyond a few hours of tinkering.

Mike Brown recalled an evening when he and his colleagues were preparing to observe on the Keck telescopes atop Mauna Kea. At Keck, astronomers observe remotely from a control room set up closer to sea level in Waimea, near the northern tip of Hawaii's Big Island, and communicate with the telescope and operator on Mauna Kea via a video link. The evening before Mike's run—the latest in a long string of observations—he had poked his head into the control room to say hello just in time to hear a tremendous crashing sound coming over the video link from the summit.

As it turned out, the clamshell-style shutter on the telescope had suffered a serious failure, tumbling to the ground and sending wires and bits of electronics flying. It luckily hadn't damaged the telescope, but the day crew was stunned and immediately set to work on trying to repair the dome. It was clearly a massive mechanical failure, and Mike was informed that he and his team couldn't possibly observe. After all, they couldn't even close the dome at the moment, and with the entire crew working on the shutter, there'd be no way to prepare for his observations.

Mike and his team, exhausted from a long stretch of observing and ready to go home, were willing enough to accept the news even if it meant their carefully laid observing plans were doomed. They changed their plane tickets, made the hour-long drive down the west coast of the island to the town of Kona, returned their rental car at the airport, and decided to make the best of things by hitting up a favorite pizza and beer joint in town. This was a treat typically reserved for the end of observing runs; astronomers widely agree that it's a solidly terrible idea to drink before running a telescope.

Several hours of drowning their sorrows later, the group had taken a cab back to the airport and started checking in for their flight home when a message arrived for Mike. It was the Keck crew. They'd determined conclusively that they couldn't possibly fix the dome shutter that afternoon, so instead, they'd gotten a head start on some observing preparations and invited Mike and his team to come back and observe…that night. Dazed and startled but knowing they couldn't possibly pass up a night at Keck, the group summoned another taxi and sobered up during the hour-long ride back to Waimea before proceeding to observe all night and get beautiful data.

Still, sometimes a malfunction can pretty unambiguously end an evening of observing. One astronomer, observing at the Cerro Tololo four-meter telescope, noticed to his surprise that the seeing—the image quality—was getting worse. The air over the telescope had slowly but surely started to shimmer and wobble, reducing the quality of his images. He was initially baffled as to why it was happening; the night had been pristine earlier, and seeing typically didn't degrade this quickly or this rapidly. Oddly, he'd gone from a beautiful crisp, cold night to the sort of image quality you see over a hot summer road or over some other sort of heat source.

When he headed downstairs, intending to stick his head out the door to see what on earth was happening, the heat source quickly became apparent: the walls of the telescope were on fire.

There had been a leak in one of the telescope instruments, sending glycol pouring through the walls, and a spark in the interior wiring had set the flammable liquid—and the walls themselves—ablaze. Apparently, the observer rather matter-of-factly grabbed a fire extinguisher and put the fire out. His colleagues suspected that had observatory management not intervened, he would have likely just gone back upstairs and kept observing; after all, he'd fixed the problem with the seeing.

o o o

Astronomy has inspired some truly innovative solutions in the name of getting things working again by whatever means possible.

Sometimes a workaround to a technical problem or broken piece of equipment can be assembled from whatever's at hand. Enterprising observers have fashioned optical light splitters out of plastic wrap from a night lunch sandwich, built adjustable spectrograph parts out of a pair of razor blades, and hung ladders, weights, or themselves from the back of a telescope's mount to serve as extra ballast and help turn the thing.

Another ad hoc solution can sometimes come up when a telescope is, oddly, too large.

There are two primary advantages to a telescope having a bigger mirror: it gives you a bigger area to collect light and the ability to produce a sharper image. The light-gathering power is usually a major perk in astronomy, since larger mirrors let us see dimmer and farther-away objects, but it can occasionally be a problem. You can imagine what might go wrong if we tried to use an enormous eight-meter mirror to take a picture of something far too bright. It'd be similar to snapping a photo with your phone's camera pointed directly at the midday sun: you'd wind up with a saturated, useless image. In fact, some CCDs used in astronomy are delicate enough that saturating them can leave a lasting imprint on the chip, like an afterimage on your eye after looking at a bright light.

This means that astronomers—like, perhaps, a certain author and her research adviser studying bright Milky Way stars—may occasionally want the sharpness or high-quality instruments afforded to them by a big telescope but *without* the full power of the telescope's entire mirror. In cases like this, it's occasionally possible to, say, hack a hole in a big circular piece of foam board and stick the foam in front of the mirror, using it to artificially shrink the effective size of the telescope so bright stars can be observed without risking saturation. There isn't exactly a solid safety procedure in place for how to climb a telescope like a jungle gym and slap on some homemade supplies, so improvisation is sometimes required. (I have a lasting memory of scrambling up this particular telescope's support structure, armed with a large piece of foam, some duct tape, and bare feet to get better traction on the slick paint of the support structure, sussing out the best place to tape the foam that would avoid getting too close to the mirror itself and wondering just how much trouble we'd get in if we screwed anything up. Fortunately, the experiment went off without a hitch.)

At other times, there's nothing to be done, and astronomers are forced to cope with what they've got. This is one of those cases where an experienced or skilled observer can have a clear edge. Vera Rubin was working at Kitt Peak one night when a malfunction stopped the telescope in its tracks. It could still observe and capture data; it just couldn't be turned. Rather than calling it quits as many observers might have, Vera simply revised her plans on the fly, worked out a target list based on what would be passing overhead, and successfully observed for the rest of the night.

On the other hand, for every exceptionally clever person thinking up a brilliant solution at a broken telescope, there are at least as many stories of incidents where the humans are the ones *causing* the problem.

Occasional mistakes at observatories aren't particularly surprising when you think about it. You've got tired people in the middle of nowhere working with incredibly delicate equipment. In the advanced stages of sleep or oxygen deprivation, with shaky judgment and a certain "damn the torpedoes" determination to power through problems and get our data, it's fair to say that even the most cautious and skilled astronomers may not always be making the best calls or serving as the best guardians of the telescopes they're working on. It's hard to find an observer who *hasn't* had either a hair-raisingly close call or an actual "oh god, I broke the telescope" moment. These moments—and the *fear* of such moments—can be almost paralyzing for young observers working at telescopes for the first time.

During my very first night of observing, I was sitting in a control room at Kitt Peak with Phil Massey, simultaneously dying to use the telescope and terrified of screwing something up. While we worked, Phil told me a story about one of his observing runs at Cerro Tololo in Chile. While observing at the thirty-six-inch telescope back in the early days of CCD cameras, his night had been interrupted by a malfunctioning CCD. That same evening, there'd been a problem at the larger four-meter telescope, so the support crew had been dispatched to the larger telescope first, leaving the thirty-six-inch waiting. Facing the uncooperative CCD and knowing there was a problem with the electronics, Phil spotted two wires, one with a plug and one with a socket, dangling out of the instrument and decided he might as well try connecting them.

Unfortunately, rather than fixing the problem, this wound up grounding the CCD. His momentary wire connection had fried the entire telescope's detector, instantly downgrading it from a cutting-edge imaging system to a very expensive paperweight. He was mortified.

That said, there was a second part to the story. The following evening, as Phil sat watching the sunset and contemplating what seemed like the end of his career, one of the telescope crew members came up alongside him and, unprompted, started sharing the story of a $50,000 imaging tube he'd been working on in a lab one evening. The crew member had left the tube on its side on a table and turned away for a moment, and the tube had rolled off the table and smashed on the floor. Phil listened and nodded, hearing both the shared disaster story and the underlying gesture of kindness that came with it.

The crew member was clearly still a valuable observatory employee, Phil went on to a long and distinguished career, and he then passed the story on to me during my first night at Kitt Peak for similarly kind reasons.

New observers can frequently terrify themselves with the mere thought of breathing wrong in a control room and destroying some enormous and irreplaceable piece of the telescope. While nobody is exactly *happy* when these sorts of things happen (the fried CCD story beat Phil back to his boss, and he was summarily chewed out after returning from his observing run), the story sharing seems to serve as a sort of simultaneous lesson and reassurance, especially for nervous young scientists: be careful, and pay attention, but understand that these things can happen, and they can happen to anyone.

o o o

When I interviewed my fellow astronomers for this book, I always asked each person if they had any favorite observing stories they'd heard second- or third- or fifth-hand. I wasn't, I explained, asking them to give me detailed or personally confirmed accounts; I was interested in the stories that had made it into the legendarium of the field, the astronomers' versions of tall tales so epic that they'd evolved over tellings and retellings.

The most common response by far was, "Do you have the one about the telescope that got shot?"

The story of the telescope that got shot has been told and retold in the fifty years since the incident actually happened, with the slow creep of embellishments that always accompanies stories like this. Still, a few basic facts are universally agreed upon and easily provable.

The incident happened in Texas (although every Texan who hears the story is quick to point out that the actual gunman was *not* from Texas but a recent transplant from Ohio) on February 5, 1970. The victim was the 107-inch telescope at McDonald Observatory, a facility nestled in a remote corner of west Texas that had begun operations less than a year prior. A major point of contention in retellings is whether the shots were fired during the day or at night while observations were happening, but a quick look through records of the incident confirms that it happened during an observing night, shortly before midnight.

The identity of the gunman also morphs between stories. Depending on who you ask, it was a crazed astronomer, a disgruntled graduate student, or, in one particularly spectacular version, a spurned lover seeking revenge for an affair. In truth, the shooter was a newly hired observatory employee, described by his superiors that evening as inebriated and suffering a mental breakdown. Whatever the underlying cause, the end result was clear: he became hell-bent on destroying the 107-inch primary mirror, entering the telescope and drawing a 9mm pistol on the operator to demand they lower the telescope until the mirror was within view. He then fired seven shots directly into the primary mirror.

However, the 107-inch mirror was nearly four tons of fused silicate glass, more than a foot thick. Rather than the hoped-for result of a shattered mirror and a destroyed telescope, the bullets simply *thunked* into the glass and stopped dead, embedded in neat little bullet holes like darts in a dartboard.

Underwhelmed by this result, the man tossed the gun aside and went after the mirror with a hammer. Fortunately, at this point, he was subdued, and the sheriff was called. Still, the chaos of the event wasn't quite over. Upon his arrival, the sheriff peered into the telescope to assess the damage and reported with horror that the mirror had been destroyed: after all, there was a giant hole in the center! Fortunately, the hole causing the panic was merely the typical opening cut in most primary mirrors to let light from the secondary bounce back to the Cassegrain focus. (I've personally always wondered what the sheriff thought could have caused such an enormous and perfectly round hole, since the gunman was shooting a 9mm handgun and not a bazooka.) Word of the Texan telescope destroyed in a shootout spread to the point where it made the national evening news. Walter Cronkite gravely reported on the catastrophic damage next to a photograph of the wrong telescope, displayed upside down.

Facing a panicked astronomy community that thought this beautiful new telescope had been shot to death, observatory director Harlan J. Smith quickly released a report summarizing the incident and emphasizing that the telescope was, in fact, fine: "The harm suffered by the mirror from his bullets and his several preliminary blows with a hammer was extraordinarily small. The damage is limited to small craters about 3 to 5 cm in radius, which reduce the light collecting efficiency by about 1 percent... The telescope resumed its observing program the following night, producing some of the best photographs...so far obtained with this instrument in its first year of use."[16]

The net result of the whole ordeal, people joked, was that the 107-inch had effectively become a 106-inch. To this day, the bullet holes are still visible in the telescope's mirror.

o o o

Even when not being held at gunpoint, it's worth remembering that an observatory can be a dangerous place. Most telescope sites combine treacherous roads, high altitudes, and remote locations several hours from medical care with heavy moving equipment and tired and fragile humans.

Altitude alone can be hazardous. At many thousands of feet above sea level, astronomers often grapple with the physical effects of the thin air. Splitting headaches, dizziness, exhaustion, and impaired judgment are all unwelcome symptoms in the midst of trying to perform challenging scientific research. More than one observer has gotten puzzled over physics concepts they'd normally know backward and forward or doggedly taken notes during a high-altitude observing run only to reread them at sea level and realize they're borderline gibberish.

The Subaru telescope on Mauna Kea sits at an altitude of nearly fourteen thousand feet. I was lucky enough during my observing runs there to never suffer from serious altitude problems, but my fellow observers and I

amused ourselves by playing with the oxygen saturation monitor the telescope kept on-site, clipping it to our finger repeatedly during the night and
watching our blood oxygen levels slowly go down. (The fact that we found
this entertaining suggests we were, at the very least, suffering the standard
giddiness that comes with mild oxygen deprivation.) Mauna Kea, one of the
highest-altitude observatories in the world, is particularly cautious about
keeping its observers safe at altitude: the observatory dormitories sit at only
nine thousand feet, a high but comfortable altitude for people to sleep at,
and people arriving at the mountain are required to spend a full night accli-
matizing there before heading up to work at the telescopes. Oxygen tanks
and breathing masks are kept in most domes if someone begins suffering
severe ill effects. Ironically, oxygen deprivation can even impact people's
visual acuity, making it harder to see small or faint objects like the stars
in the dazzling skies over these mountains. Several people told me about
an interesting but decidedly off-label use of oxygen tanks as a result: stepping outside, looking up to what seems like a solidly "meh" night sky to an
oxygen-starved brain, and then taking a hit of oxygen and watching the stars
bloom into view.

It's also worth remembering that an observatory summit is hours from
medical care. In the event of a significant medical emergency, there's little
to be done beyond rushing back to civilization. Actor Alan Alda once visited Las Campanas Observatory in Chile to see an observatory in action
and interview astronomers there for an episode of his PBS show, *Scientific
American Frontiers*. While on the summit, he suffered an excruciatingly
painful intestinal strangulation that required surgery and had to take a panicked ambulance ride from the remote observatory summit to the hospital
in La Serena, a story he relates in his book *Never Have Your Dog Stuffed: And
Other Things I've Learned*.

The Chilean observatories of the Atacama Desert are particularly well-
equipped to deal with medical emergencies, in part because northern Chile
is no stranger to physically dangerous jobs carried out in places even harder

to reach than mountain summits; the infamous San José Mine, where thirty-three miners were trapped underground for sixty-nine days in 2010, is also in the Atacama region. Chilean law requires that a medical professional be nearby at these remote sites in case of emergencies. One astronomer also recalled a colleague who, back in the days of in-dome observing, had gotten engrossed in his observations at Cerro Tololo, removed a safety chain that was in his way, and fell off a platform, landing flat on his back on concrete. Concerned colleagues insisted he go see a doctor, an old man who sat chain-smoking on a stool and instructed the astronomer to "walk for me" up and down the hall. After a short walk, the doctor—correctly—declared him fine and explained he'd worked in mines for forty years and seen so many fall injuries that he could determine just from the astronomer's gait that he hadn't been badly hurt.

Falls themselves are a hazard in astronomy, with a particularly rich history from the era when scrambling up ladders, into prime focus cages, or onto high platforms was the norm. Astronomers walking around Cassegrain platforms or interior dome catwalks in the dark have chalked up a fair number of broken legs or fractured backs thanks to falls. In almost every one of these stories, there's also some variation on the injured astronomer calling plaintive instructions to their observing companion—"Finish this exposure! And then point to the next target!"—as they're carted away for medical attention, in pain but still unwilling to lose so much as a moment of telescope time.

The darkness of the dome also doesn't help matters: astronomers have been bruised, severely concussed, and even knocked out cold by excitedly dashing from the warm room into a pitch-dark dome and running head-long into a telescope counterweight or concrete mount. The freezing temperatures also caused problems back during the era of guiding telescopes by hand and working out in the domes. More than one astronomer pressed their eye to the eyepiece for a long exposure and then discovered that they'd frozen the tears and tender skin around their eye directly to the metal.

Apparently, one observer managed to unscrew the eyepiece, carried it into a nearby warmer room, and then calmly waited for it to defrost so they could safely detach and go back to observing.

o o o

It's tempting to hear stories like this—often delivered with a wry chuckle, especially by embarrassed observers remembering their own self-inflicted injuries—and start thinking of these situations as a kind of amusing slap-stick comedy. It's not hard to picture goofy-tired, altitude-giddy scientists careening around domes and clocking their heads on protruding bits of equipment like a *The Three Stooges at the Telescope* skit. However, the truth is something much, much more serious.

The occasional injury may seem like a one-off incident, but in truth, each one is a reminder of how precarious observers' positions can some-times be. Even with care taken to set up safety protocols and buddy systems to protect people from harm, the fact remains that astronomers are, for the sake of our profession and scientific passion, working in the dark inside massive moving buildings with heavy equipment that requires people with professional training to operate it with extreme care.

Nowhere is this sobering reality more keenly felt than in the story of Marc Aaronson.

Marc Aaronson was an astronomer at the University of Arizona study-ing one of the most exciting and enduring questions in astronomy: the value of the Hubble constant. Originally proposed by Edwin Hubble him-self in 1929, the constant is a single number—a ratio of speed divided by distance—slotted into a deceptively simple equation that's meant to accu-rately describe the expansion of the universe. Measuring the number itself, however, is devilishly difficult and depends on developing tricks for *per-fectly* determining distances to galaxies that are billions of light-years away. As a result, the exact value of the Hubble constant (which bounced between

50 and 100 over the years and has currently been narrowed down to somewhere between 65 and 75 depending on who you ask) has been a source of absolutely ferocious debate among astronomers for almost a century.

By the 1980s, Marc was one of the exciting young minds working at the leading edge of this field, a fixture at the key conferences where new research related to the Hubble constant was presented and debated. He was also an extremely skilled and enthusiastic observer, with large telescope experience dating back to his undergraduate years at Caltech. By age thirty-six, he had already received several prestigious research awards, including Arizona's George Van Biesbroeck Award, Harvard's Bart Bok Prize, and the Newton Lacy Pierce Prize awarded by the American Astronomical Society for outstanding achievements in observational astronomy.

On the night of April 30, 1987, Marc was observing with a student at the Kitt Peak four-meter telescope, part of his ongoing work to determine the distances to other galaxies and refine the Hubble constant. Early in the night, he asked that the telescope and dome be turned to point at a new galaxy and then hurried to the dome catwalk to get a good look at the sky conditions.

Like most telescopes, the four-meter consisted of a tall cylindrical building with a dome nested on top of it. When moving between objects, the entire dome would turn along with the telescope inside to keep the open portion aligned with the telescope's view. The telescope's exterior catwalk sat just below the height of the turning portion of the dome, with a portal-like door set about knee-high in the wall to connect the catwalk and the dome interior. Several parts of the giant dome shutter were actually low enough to hit this door. As a result, there was an interlock system in place to prevent the door from being opened while the motors were running to turn the dome. However, a feature nobody had fully considered was that when the motors switched off, the dome didn't quite stop cold. Weighing more than five hundred tons and moving at almost a foot per second, it would silently coast for several feet before stopping.

That night, Marc and his student arrived at the catwalk door, ready to head outside. As soon as the motors stopped, Marc opened the door to the exterior catwalk, unaware that the dome was still coasting. Just as he stepped through the portal, part of the coasting shutter struck the door and forced it shut. He was killed instantly.

Marc's death shocked and devastated the entire astronomy community. It was a horrific accident, and the field had lost a great friend and talented colleague at the height of his career. The "Large Scale Structures of the Universe" research conference held two months later was dedicated to his memory. In a paper from the conference that summarizes Marc's work and final research (one of several publications listing him posthumously in the honored position of lead author), his colleague Ed Olszewski wrote, "we need more scientists like Marc Aaronson, with lots of ideas, and the energy to see them to fruition."[17]

Today, after a full safety audit of the telescopes on the summit, the four-meter telescope has a triple interlock system in place at the dome. Many other observatories also revised their own safety procedures and systems after the accident. The University of Arizona now awards the Aaronson Lectureship every year in Marc's honor to an individual who "has produced a body of work in observational astronomy which has resulted in a significant deepening of our understanding of the universe."[18]

CHAPTER SIX

A MOUNTAIN OF ONE'S OWN

"Could you find the observer for me?"

I glanced up from the computer screen, confused by the question and whether the person speaking was talking to me. True, as the only person in the room—the Waimea remote observing room set up for astronomers using the Keck telescopes on Mauna Kea—odds were good the guy who had just walked in had meant his question for me, but I also *was* the observer. I was busy with another night of observations for my PhD thesis, juggling scattered clouds, an ornery infrared spectrograph, and a target list that I was rearranging on the fly.

My targets were more galaxies that had hosted mysterious dying stars emitting gamma rays, but tonight, they were particularly far away, so distant that the expansion of the universe was whisking them away from me at close to the speed of light. At that breakneck speed, the very light emitted by these galaxies was getting distorted by the effects of relativity and producing a phenomenon known as redshift, the electromagnetic cousin of the Doppler effect. The Doppler effect is a classic one: if a car leans on its horn as it approaches you, the pitch of the horn creeps higher and higher, but as the car zooms past you, the pitch appears to plunge, getting lower and lower as the car drives away. The effect comes from sound waves being compressed or stretched relative to you as the listener. As the car comes toward you, the sound waves get compressed and are a little bit shorter when they arrive at your ear, so they sound higher, while when the car is driving away, the sound waves get stretched and are a little bit longer when they arrive at your ear, so they sound lower.

The same thing was happening to my galaxies on a light-speed scale. For sound, shorter waves sound higher, and longer waves sound lower. For electromagnetic waves—light—shorter waves look bluer, and longer waves look redder. These galaxies were moving away from us so fast, screaming along as the universe inexorably expanded, that the very light they emitted was being stretched. Light that would be visible to the naked eye if I traveled *to* the galaxy and looked directly at it was being stretched so drastically that it was arriving at Earth as infrared light, a wavelength too long for the human eye to see and one that can only be detected by telescopes using instruments specifically designed for infrared light.

This was the data I hoped to capture, but I knew from experience that the infrared instrument I was using could be temperamental. The clouds meant I'd also already lost too much time in the night to make it through my whole galaxy list, and I was trying to organize the hours I had left. Did I go for the brightest and easiest targets or prioritize the fainter, farther, but more exciting galaxies? To answer the question, I needed to recalculate when each galaxy would be at the optimum spot in the sky for observations, decide which galaxies I could conceivably put off until my next observing night in a month, and crunch tonight's data as it came in so I could make sure the instrument was working properly.

So yes, I was the observer. I was also pretty busy. Both of these things seemed fairly self-evident.

"I'm the observer. What's up?" I was too preoccupied to fully turn off "brusque New Englander" mode, but I at least remembered to look up and smile. The guy who'd spoken seemed to be another astronomer:

middle-aged, with a laptop backpack, a T-shirt emblazoned with the name of another observatory, and a canvas bag bursting with night lunch snacks. As another astronomer, he might be new to Keck, or a veteran looking to say hello to a colleague, or coming in with a question about the telescope or observatory or weather that the current observer would be best equipped to answer.

The guy paused. "No, I mean the person in charge."

That…was still me. I looked down to check on how much time my current exposure had left and tapped a couple of commands to continue processing my data. "How do you mean?" I thought he might be looking for someone who worked at Keck, maybe a member of the support staff or maintenance crew.

Now he looked a little impatient. "Whose telescope time is this?"

"Oh, mine. I'm Emily. I'm at the University o—"

"No, I mean who's the PI that you're working with? I didn't recognize his name on the schedule…"

Ah.

PI was short for principal investigator. In academia, this is the person who leads a proposal for funding or telescope access and is officially granted the money or time. Being a PI at Keck means a certain amount of knowledge and experience and a degree of seniority. My visitor clearly didn't think that I looked like PI material.

It was weird, to be sure. I was the only person in the room, sitting directly in front of the main bank of computer screens, surrounded by sky charts and a laptop and a gigantic four-inch binder holding complete dossiers for every galaxy I planned to include in my thesis. Just as the man had walked in, I'd been talking over the teleconference system with the operator on the mountain, telling them my plans for what target I'd be observing next. What else could I possibly look like?

That said, I knew what I looked like. Of course I did. I was in my fourth and final year of grad school (officially experienced, in graduate student years), which in my case meant I was twenty-five. I was dressed for a long, comfortable night of staying warm and observing, which meant baggy flannel pajama pants, a long-sleeved T-shirt with cartoon penguins on it, and a striped wool hat. I was sporting pigtail braids; after a lifetime of pixie cuts, I'd sworn I wouldn't cut my hair until I got my PhD, and four years in, my hair had finally gotten long enough for a friend to teach me how to braid. I had a bag of Goldfish crackers and a bag of peanut butter M&M's perched next to me, my snacks of choice for observing runs. I was sitting in one of the giant built-for-tall-people office chairs with my legs tucked under me; I'd kicked off my shoes, and my socks were bright yellow and had smiley faces on them. I looked, in short, like a girl.

I didn't look like the PI.

"I'm the PI. Emily Levesque. I'm a grad student at the University of Hawaii."

"Oh." He nodded, pursed his lips. "I'm on tomorrow night." He paused.

I waited; this was usually the point where two astronomers would start chatting about what they were studying, how the telescope was doing, how the weather was looking.

"I'm from Caltech. We're studying..." He trailed off. "Anyway, yeah, I just had a question for... I'll find someone." He was halfway out of the room before stepping back in. "How's it going?"

"Okay! Some clouds, but the fog from last night cleared, so we've been open all night. Seeing's pretty good, getting better..."

"Cool. Later."

I got back to work—the exposure ended, we moved the telescope, I double-checked the clouds, and the next exposure started while I grabbed the new data—but a persistent little worm of doubt was wiggling around in the back of my brain. I suddenly wondered if I should put my shoes back on, go get some jeans, or tuck the braids away. Was I really acting professionally? Was this really what a PI looked like?

Would that conversation have gone the same way if I had been a man?

In a way, it was the ambiguity that bugged me the most. There's rarely a sign that lights up for women and blinks out THIS IS SEXISM. THIS PERSON IS BEING SEXIST RIGHT NOW. I was sure I could think of plenty of other explanations for what had just happened. Maybe the guy was just awkward or shy, even though he'd been the one to start the conversation and had clearly wanted to talk to *some*one. I also looked young. I *was* young; twenty-five is a pretty early age to be at the helm of one of the largest telescopes in the world. Maybe he assumed I wasn't a PI because of my age. Then again, there'd been a young male professor at MIT widely praised as a superge-nius; his age hadn't seemed to bother anyone. Maybe it was the cartoon penguin shirt; that didn't exactly scream gravitas, although plenty of the guys I knew—hell, plenty of the professors—wore Super Mario or Teenage Mutant Ninja Turtles T-shirts to work, and it never seemed to disqualify them from being in charge. I'd been sitting cross-legged in the chair like a little kid, which must have looked silly. At five-foot-two, all the chairs in the room were too big for me, and I was trying to stay comfortable during my ten-hour night. Goldfish as a snack don't really communicate maturity, do they? Were my socks too goofy? Would he have refused to believe I was the PI if I'd been a guy? Did it matter?

I already got asked with surprising frequency—often by strangers, when I told them what I studied or where I went to school—if it was hard being a woman in a male-dominated field. I always quickly answered that it wasn't. I meant it. I really never thought about it.

Maybe it was the braids.

o o o

If I'd been an astronomer fifty years prior, I wouldn't have been struggling with the pernicious "is it or isn't it?" question when it came to sexism. The answer would have been a solid yes.

As late as the 1960s, the Monasteries at Mount Wilson and Palomar in California had strict rules against allowing women to sleep there, and women could not officially apply for time or serve as a PI (ahem). As in so many other science fields, women were studying astronomy and teaching and working in the field with great ardor and enthusiasm, but the attitude that working at a telescope was a man's job still prevailed. Unsurprisingly, none of these rules were wholly effective at *banning* women from the telescope; they just ensured that extra challenges were put in their way to make things a bit harder. Barbara Cherry Schwarzschild was just one example of this; while her husband, Martin, was a brilliant astronomer, Barbara was the one who knew her way around the telescope. The time was always granted in Martin's name, but Barbara was inevitably the one doing the observing.

Margaret Burbidge began her observing career in a similar fashion. In 1955, Mount Wilson officially began granting time to her husband, Geoff, a brilliant theoretical astrophysicist. However, more than one fellow observer joked that at the time, Geoff didn't know one end of a telescope from the other. The real situation was an open secret on the mountain: Geoff's telescope privileges enabled Margaret to observe as his "assistant." She went on to perform groundbreaking research: in a research paper that she led along with Geoff, William Fowler, and Fred Hoyle, she put forward the theory that all but the lightest chemical elements are produced in stars. In short, she was one of the scientists behind the famous adage "we are made of star stuff." Still, during their observations, Margaret and Geoff were forced to stay in a small cottage on the observatory grounds because she was banned from sleeping in the Monastery.

In 1966, Ann Boesgaard was the first woman granted time under her own name at the Mount Wilson 100-inch, but she too was relegated to the cottage. Around this same time, Elizabeth Griffin began visiting Mount Wilson from Cambridge University with her then-husband, Roger. This in itself was a feat; she and Roger were both astronomers but worked in different research areas, and the research council in the UK allowed him to have a grant but not her. His grant requested funds for both of them to travel to Mount Wilson for their research—a long and expensive trip—but this point was argued over for weeks and ultimately brought to the Astronomer Royal (an official post in the Royal Household of the UK), asking whether the grant should be allowed to fund both scientists' travel expenses. The Astronomer Royal wrote back asking, in essence, "I understand why Dr. Roger Griffin needs to go, but why does Mrs. Griffin need to go too?"[19] In spite of the administrative challenges, the Griffins won in the end, and both were granted funds to visit Mount Wilson. However, this meant that between Elizabeth and Ann, they now had *two* women on the mountain, a truly vexing logistical challenge to the astrophysicists in charge, as they still wouldn't allow women in the Monastery.

The first female observers at these telescopes got thrown some extra hurdles thanks to the housing policies. Staying in a lovely little mountain cottage rather than a spartan-sounding Monastery might sound like a treat. In reality, the cottage at Mount Wilson was rudimentary and had electricity but no running water, no shower or bath, and only an ornery woodstove nicknamed "Old Dudley" for heat, making it a frigid place to stay during the winter months. Ann and Elizabeth both recalled the misery of coming back to the cottage after a long winter night of observing and having to build a fire in Old Dudley to get the place warm enough to sleep while the male astronomers could happily go straight to bed in the heated Monastery. The cottage was also situated out near Echo Point, where tourists and hikers would wander up and shout cheerfully to each other during the day when observers were trying to rest.

Bathrooms were another perennial source of supposed stress in these situations; where, the powers that be wondered, would the women pee? This had long been the argument at Palomar: women couldn't observe at the 200-inch because there was no toilet for them to use. (Recall where the *men* were peeing while they were stuck in the prime focus cage all night with empty dry ice thermoses at hand...) The same fuss was kicked up over the dormitories. When Anneila Sargent and Jill Knapp became the first two women to stay in the Monastery at Mount Wilson, there was much worry over bathroom arrangements. Observatory staff fretted about these poor women being forced to share a bathroom with male astronomers. This concern came in spite of the fact that both of them were, in fact, *married* to male astronomers; as Anneila pointed out dryly, "we'd had this experience."[20]

In 1965, Vera Rubin became the first woman to receive time at the largest of the California telescopes, the 200-inch at Palomar. By then, she was permitted to stay in Palomar's Monastery, but the observatory apparently still had to work out a few kinks. She was taken on a tour of the 200-inch telescope grounds on her first, cloudy night and shown the "famous toilet," a single-occupant room with a prominent—and amusingly redundant—sign on the door labeled MEN. Vera described her simple fix to this problem in "An Interesting Voyage," a professional memoir that she wrote for the *Annual Review of Astronomy and Astrophysics* in 2011: "On my next observing run, I drew a skirted woman, and pasted her up on the door."[21] Solved.

Vera went on to make one of the most incredible discoveries of observational astronomy when she began studying the motion of spiral galaxies in 1967. Astronomers at the time expected the wispy outer arms of the spirals to be spinning slowly through space, while the inner regions near the galaxy's center would move much faster. Their reasoning made sense: everyone could plainly see that the mass in these galaxies—stars, dust, and gas—was visibly clustered in their centers, and the laws of gravity dictated that anything farther away from this central bundle of mass would feel a weaker gravitational pull and rotate more slowly.

Vera expected to confirm these predictions, but she got a surprise: the outskirts of these galaxies weren't moving slowly at all. Instead, the gas and stars at a galaxy's edge seemed to move just as fast as the gas closer to the center. No matter how many galaxies Vera studied—and she observed dozens—each showed the same odd behavior. After puzzling over the results for months, Vera realized that her data could be explained perfectly if each galaxy had, in addition to its visible stars, gas, and dust, an invisible halo of matter contributing to its mass. She had identified the first observational evidence of dark matter.

Today, we still don't know what dark matter actually *is*, but we certainly know it exists. After Vera's discovery, dark matter became a linchpin of how astronomers explain the history and evolution of the cosmos, and new observations continued to confirm immense quantities of invisible mass in the universe. The discovery spawned entire new subdisciplines of physics, and Vera went on to eventually win just about every prestigious prize offered to a professional astronomer (with the exception of the Nobel in Physics, which has long had a sizeable blind spot for groundbreaking research done by women). During most of her research on dark matter, she remained one of the only women working at the observatories that granted her time.

Some years later, Vera was observing with a colleague, Deidre Hunter, at Las Campanas Observatory on an evening when, as it turned out, all the astronomers working on the mountain were women. Vera gathered the whole group into one of the control rooms for a photograph to commemorate the event. Elizabeth Griffin recalled a similar night at Mount Wilson. While observing on the 100-inch along with a colleague of hers, Jean Mueller, and a woman serving as the night assistant, the three of them collectively realized that due to a quirk of the math, they were the only astronomers working on the mountain, making Mount Wilson an all-female observatory for the evening. Both of these nights happened in 1984, the year I was born.

o o o

I grew up in the grrrl-power you-can-have-it-all mentality of the 1990s. My family and teachers had sent the clear and unambiguous message that I should pursue whatever goal I set my mind to, regardless of gender. In my boyfriend, Dave, I'd found the strong and supportive partner who had seemed so elusive in some of the career woman movies of my childhood.

Katy Garmany, Deidre Hunter, and Vera Rubin in the control room of the four-meter telescope at Kitt Peak National Observatory in January 2000; like the photo Deidre described from 1984, the picture was taken at Vera's request to mark the occasion of three women astronomers in the control room.
Credit © John Glaspey.

He made it clear that we came to our relationship as equals, that we would support and value each other's careers and challenge ourselves to always dream bigger.

When I first entered the field, it was all too easy to convince myself that gender must have very little to do with astronomy.

When I first heard stories of Ann and Vera and their contemporaries, my brain reflexively placed them in ancient history. In college and even as a graduate student, whenever I heard a story about "those women who were forbidden from staying in dormitories and observing at telescopes," my mind conjured up imagined pictures that were some weird sepia-toned amalgam of *Anne of Green Gables* and suffragette documentaries—men with pocket watches, women with long, black Edwardian skirts—because surely *those* were the quaint ancient days when women were told they couldn't do things just because they were ladies. It was all too easy to forget that these stories had happened in the 1960s, in the era of color photos and jeans, of hippies and civil rights, that most of these women were born after my own grandmothers. I'd met almost all of them, and Ann Boesgaard was my first research mentor at the University of Hawaii.

It's hard to say whether the pace of change that brought astronomy from Vera Rubin being the first female PI at Palomar in 1965 to all-women observing nights in 1984 has slowed down, sped up, or stayed constant. Astronomy is undeniably different today; according to the American Institute of Physics, women earned 40 percent of the 186 astronomy PhDs awarded in 2017.[22] However, the improvement in gender representation is still starkly absent along other axes. Women may have represented 40 percent of the doctoral degrees awarded in 2017, but Hispanic women comprised only 4 percent, and African American women made up 2 percent.[23] The 2007 Nelson Diversity Survey found that among the top forty astronomy departments in the United States, only 1 percent of professors of any gender were Black and another 1 percent were Hispanic. Representation is improving for current astronomy undergraduate and graduate students (helped by programs such as the Banneker Institute at Harvard, named for Benjamin Banneker, the first African American astronomer), but these numbers are still painfully small. There's also no current data quantifying how many of these astronomers are observers (as opposed to scientists who work in theoretical astrophysics or other subfields), despite a rich history of observing among astronomers of color. Harvey Washington Banks, the first African American to earn a PhD in astronomy in 1961, was an observer who specialized in spectroscopy and precise orbital measurements. He was followed by Benjamin Franklin Peery in 1962, Gibor Basri in 1979, and Barbara Williams in 1982, the first African American woman to get a PhD in astronomy. Like many observers, other prominent Black astronomers came to telescopes through backgrounds in physics and engineering. These included Arthur B. C. Walker II, a pioneer of rocket-based and X-ray astronomy, and George Carruthers, the father of an entire branch of astronomy thanks to his invention of cameras and spectrographs that could detect ultraviolet light.

Perhaps most importantly, it's continually becoming less radical to demand change and prioritize equity and inclusion across the field, including when it comes to observing and telescope access. The Hubble Space Telescope recently switched to a dual-anonymous system in its yearly review of observing proposals after an internal study found that success rates were consistently lower for proposals led by women than those led by men; after removing names from the proposals, the percentile difference between genders disappeared.

At the same time, even several decades later, my colleagues and I are still largely unsurprised to be the only women in a room or on the mountain, to the point where it barely even registers. I can easily remember observing runs where I was the only woman at dinner but cannot recall a night where, like Deidre or Elizabeth, I was on a mountain of only women. Part of this can be blamed on the small size of our field, the simple effects of statistics, and the fact that powerful senior positions are occupied by senior people who represent the field's demographics from a few decades ago. Still, the all-women observing nights in 1984 were also quirks of statistics. Around the same time, a running joke at Las Campanas Observatory claimed that there was "a woman behind every tree"…on a treeless summit.

o o o

Observatories, of course, are no more immune to issues of sexism or racism than anywhere else in the world. Several women I interviewed told me they had been harassed or assaulted at telescopes. Others described scenarios on mountains that, while not outright assault, still made for a deeply discomfiting environment. Two different women—who weren't aware of each other when I spoke to them—described working at the same observatory with a telescope operator who would openly watch pornography on his laptop as his source of entertainment in between telescope moves. A few others described instrument shops or maintenance areas decorated with naked

pinup calendars and Playboy centerfolds. (Several men also mentioned seeing these and speaking up to object.)

Some women shared ironic stories of sexism at observatories stemming from the attitude that women must surely need to be protected while on the mountain. It's obviously a fair sentiment in moderation—any decent person wants to ensure their colleagues are safe—but when tied up in gender, it can quickly cross the line into patronizing. In 2010, one colleague of mine was enjoying her first night as the PI of a small telescope in Arizona when she received a phone call from another astronomer, demanding that she immediately stop observing because he didn't think it was safe for her to be up at the telescope by herself as a woman. With the telescope sitting idle, the astronomer on the phone insisted she bring in a male chaperone before continuing her work. Other women described observing while pregnant, but rather than physical discomfort or medical concerns, they mainly recalled freaked-out male colleagues who insisted they carry walkie-talkies with them at all times or not lift anything.

I asked Dara Norman, an observer who studies distant galaxies with supermassive black holes at their centers, how she thought race had impacted her experiences as an observer. She recalled her arrival at telescopes in Chile being met with genuine surprise by observatory staff who weren't expecting the incoming astronomer from the United States to be a Black woman. John Johnson, whose work focuses on the discovery of exoplanets—planets beyond our own solar system orbiting other stars—has had similar experiences. The stark demographics of being one of only a couple dozen Black astronomers are a constant reality, and at observatories, John would occasionally meet colleagues or telescope staff whose reactions ranged from mild surprise to blunt "are you in the right place?" skepticism (though he pointed out that this sort of reaction is all too familiar for scientists and academics of color wherever they go).

That said, almost everyone described working at the telescope as a refreshing departure from these sorts of pernicious issues that were so often

a part of everyday life. With the technical challenges and chaos and number of things that can go wrong, there's a certain camaraderie that everyone can quickly come to share at a mountain summit. Often, the collection of observers on the mountain becomes a community of scientists, with gender and race taking a back seat to bonding with your colleagues and buckling down for the work that everyone was there to enjoy.

Even women who had experienced harassment at observatories in the past were quick to point out that this was the fault of the men doing the harassing and those who had permitted those men to infringe upon the mountain. As scientists, they still loved observing and still thought telescopes were wonderful places to work.

Multiple women mentioned the pleasant fact that while observing, they weren't the only girl in class or the only female professor or similar. They were simply a person on the mountain, with whatever smattering of colleagues happened to be there at the time, immersing themselves in astronomy all day and battling weather and finicky instruments all night. These were the experiences that resonated the most with me as a female observer and a woman in astronomy. A telescope can be a place of wonderful quiet, of walking comfortably alone through the dark and night and getting to soak up the beauty of the sky as a scientist and human. If, as Virginia Woolf pointed out, a room of one's own is required in order for women to write and create beautiful stories, it's thrilling to imagine what can happen on clear nights at telescopes with a mountain of one's own.

o o o

It might be tempting to imagine that the world of scientific research can exist apart from mere human concerns, that lofty pursuits like exploring the heavens can simply brush aside such seemingly small issues as gender, race, or other sources of personal conflict in pursuit of some pure scientific truth. In reality, the exact opposite is true. These issues are anything but small for

the countless human scientists who live with them, and tackling them as a community is a vital part of our fundamental roles as scientists and citizens. The challenges that come with controversy and conflict have also long held a prominent place in astronomy.

When people think of "controversy in astronomy," it's easiest to focus on the ideological arguments of history: Galileo vs. the church, the age of the universe pitted against various religious creation myths, or the churning arguments of UFO skeptics. On the last point, I'll say only this: after fifteen years of observing and talking to other astronomers about the weirdest things in the universe and their most incredible experiences while observing, I've never spotted a UFO and never once had someone report a sighting. That said, even professional astronomers are occasionally taken aback by things like the brilliant glow of Venus, a passing satellite, or the underlit bellies of geese, all things commonly and mistakenly identified as UFOs.

It's true that the things telescopes teach us about the universe can be shocking and humbling and sometimes a source of vigorous scientific debate (consider the Hubble constant argument, Pluto's status as a planet, and other topics that astronomers can endlessly spar over). Still, some of astronomy's most heated conflicts in the past few decades haven't centered around the science but around the observatories themselves.

Beginning in the 1980s, groups ranging from radical environmentalists to recreational hunters and representatives of the San Carlos Apache Tribe all levied protests and legal action against telescopes being built on Arizona's Mount Graham. In the early 1990s, an observatory in South America was the subject of several lawsuits that could be traced back to a supreme decree by Chilean dictator Augusto Pinochet. In 2015, protestors blocked Mauna Kea Access Road to prevent construction equipment, there to break ground on a new telescope, from reaching the summit. Coming across sound-bite summaries of these protests in the news or on the internet, most people were, at first glance, bemused. After all, why on earth would anyone be protesting a *telescope*?

The nature of these controversies is easier to understand when you consider not the telescopes themselves but where they are built. Finding good astronomical sites is exceptionally difficult; as we've moved to push the limits of telescope technology, it's become more and more important to pair cutting-edge telescopes with the few places on Earth that are ideally matched to their abilities, and these locales can become sources of conflict in their own right. Dr. Leandra Swanner, now a professor at Arizona State University, undertook a detailed analysis of these debates in her 2013 PhD thesis, titled "Mountains of Controversy: Narrative and the Making of Contested Landscapes in Postwar American Astronomy."

Most vocal or controversial challenges to telescopes fall into one of three broad categories. The first covers environmental concerns. Observatories are built in pristine mountain areas. The environmental impact of a telescope pales in comparison to most other construction efforts on mountaintops, such as hotels and ski resorts, but large construction projects still involve blasting roads, hauling equipment, leveling ground, pouring foundations, erecting large structures, and introducing foot and vehicle traffic.

The second involves land rights, which can quickly get complicated. The place where a telescope is being built must be properly apportioned for that purpose. This typically requires getting a permit to build on the site, but determining who rightfully owns the site's land and who has the authority to issue permission for building on it is sometimes easier said than done.

Finally, observatory mountaintops can have spiritual and cultural significance to local indigenous populations. Any observatory always does better when it's far away from busy and well-lit population centers, and it becomes tempting to look around a remote mountaintop and declare it uninhabited and there for the taking. However, in some cases, these places are homes—in spirit if not in material practice—to generations of people who may feel bound to be good stewards and protect the land from outside influences.

Mount Graham in Arizona was chosen in the early 1980s as the site of a new observatory, initially imagined as a grand expanse of more than a dozen telescopes built on a mountain that boasted excellent atmospheric conditions for astronomy. As it turned out, the mountain was also an excellent place for squirrels—the Mount Graham red squirrel, to be precise, a species that was investigated and placed on the endangered species list in 1987—and a popular recreation area. The result was a decidedly odd alliance between environmentalists and local hunting clubs, both objecting to the construction of the observatory.

After years of legal disputes, the tone of the conflict had turned intensely hostile: radical environmental activists smashed equipment, cut down power lines at observatories (including some that were completely unaffiliated with the Mount Graham efforts), and lay in the road to block construction equipment. One supporter of the observatory received a death threat, and another received a dead squirrel in the mail. Observatory backers, fed up with disputes and delays, enlisted the help of Arizona senators and a powerful Washington lobbying firm to push a rider through Congress in 1988 that allowed them to begin construction immediately without fulfilling the conditions of the Endangered Species Act, a tactic that was met with outrage and made Mount Graham a national story.

In 1991, a nonprofit group founded by members of the San Carlos Apache Tribe filed another lawsuit to stop construction of the telescopes on Mount Graham, stating that the mountain was sacred and construction would desecrate traditional Apache religious sites and burial grounds. The suit was dismissed by a judge in 1992, but opposition to the telescopes on Mount Graham remained strong. At that point, the controversy encompassed environmental and ecological concerns, the value and fate of a mountain that had been declared spiritually sacred, and a growing sentiment among some activists that the observatory was, as Swanner summarized in "Mountains of Controversy," a "symbol of cultural genocide."[24]

Press coverage of the issue did no one any favors. From a media perspective, this was a fascinating story if properly simplified, pitting a tiny helpless squirrel against rapacious bulldozers making way for enormous polluting telescopes. The news also gleefully reported on some of the nuttier squirrel-based protests (such as Earth First! members dressing up as telescopes and squirrels to act out a skit and disrupt a public hearing about the observatory). While some stories did acknowledge the objections of the San Carlos Apache, most preferred to reduce the complex and multifaceted protests to "Squirrels vs. 'Scopes."[25]

Today, Mount Graham operates as a typical astronomical observatory, and its red squirrel population is doing well. Biologists carry out an annual squirrel census every year. Astronomers observing at Mount Graham must be fully briefed and sign forms agreeing that they will not kill, harm, or taunt the squirrels. Observatory proposals to offer astronomy outreach and educational grants for university programs to the San Carlos Apache Tribe were described by some members of the tribe as blatant attempts at bribery and have been rejected.

Mount Graham is one example of a mountain that, between environmental issues, land rights, and cultural objections, found itself with the full trifecta of controversies that can become tangled up with telescopes. Mauna Kea Observatory, on the Big Island of Hawaii, is another.

o o o

Mauna Kea is a dormant volcano formed only one million years ago as part of Hawaii's volcanic hot spot. Measured from the ocean floor, it's thirty-three thousand feet tall, higher than Everest. The high slopes of Mauna Kea are classified as alpine tundra. The landscape is completely bare of vegetation, with sweeping red cinder cones set against a two-toned backdrop of fluffy cotton-ball clouds below the summit and a pristine sky above it.

This environment—the ground stretching above the clouds, the crisp, bone-dry air, and the utterly perfect blue sky—is what makes Mauna Kea such an exceptional place for astronomy. The telescopes at the summit have contributed cutting-edge data to just about every subfield in astronomy, from discovering comets and asteroids to capturing light from the other side of the universe. On a single night at Mauna Kea, the telescopes might capture eagerly anticipated data for a dozen or more different research projects, making discoveries that are impossible from anywhere else on the planet.

Like other Hawaiian volcanoes, Mauna Kea is also considered a sacred mountain in the Native Hawaiian religion. With a name meaning "white mountain," deriving from its snow-capped summit, Mauna Kea is believed to be the home of Poli'ahu, one of the snow goddesses in Hawaiian mythology. The mountain is also seen as representing the umbilical cord of the Big Island, connecting the land to the sky from which it was born.

The Hawaiian cultural connection to the cosmos is powerful. Traditional Hawaiian and Polynesian wayfinders relied on, among other things, an encyclopedic and detailed knowledge of the night sky gained through oral traditions and memorized by navigators. Polynesian explorers were able to cover thousands of miles across the Pacific Ocean, traveling with only the stars to guide them.

Still, astronomy on Mauna Kea has been a persistent source of controversy since the first telescope on the mountain began operating in 1970, with objections questioning the impact of the telescopes on the mountain's environment, cultural activity, and even the view that Big Island residents enjoyed (an early concern had been that the white telescope domes dotting the summit, which can be seen from much of the island, would be ugly). A 1983 development plan and environmental impact statement laid out proposed plans for the mountain's development through the year 2000 and allowed for the construction of no more than thirteen telescopes, a limit reached in 2003.

Then, in 2009, Mauna Kea was chosen as the site of the Thirty Meter Telescope, or TMT.

One of several "Extremely Large Telescope" projects (telescopes with mirror diameters larger than twenty meters), the TMT is an international collaboration of astronomy research institutes, and in 2009, it selected

Mauna Kea as the best spot on the planet for its construction. When complete, the TMT will be the largest telescope in the Northern Hemisphere and the second-largest in the world, dwarfing the ten-meter Keck telescopes also perched on Mauna Kea. It will be capable of producing images twelve times sharper than the Hubble Space Telescope and answering questions about the earliest moments of our universe, the mysteries of black holes, and distant planets that could potentially host intelligent life. Building the TMT on Mauna Kea would cement Hawaii's already well-established place as a world leader in astronomical research.

Opponents to the telescopes on Mauna Kea were strongly against the TMT—it was seen as a clear violation of the thirteen-telescope limit promised in 1983—and began pursuing ways to halt its construction. A legal challenge to the TMT's construction permit was enough to delay the process for several years.

By 2014, the University of Hawaii had agreed to remove three existing telescopes from the observatory in an effort to comply with the thirteen-telescope promise from 1983. The TMT consortium committed millions of dollars to initiatives supporting local employment and STEM education on the Big Island. The TMT would also be as unobtrusive as possible, painted with a special coating that would reflect the sky and ground and built on a northern plateau at a lower elevation, rendering it invisible from 86 percent of the Big Island. The telescope's site had been surveyed to confirm there were no burial sites or other artifacts near its location, and monitors would remain on-site during construction with the authority to immediately halt work if any new archaeological or cultural discoveries were made. With plans in place and the legal disputes resolved, a permit was issued for construction to begin in 2015.

That April, a large group of protestors headed up the mountain and blocked the Mauna Kea Access Road to stop construction equipment from reaching the TMT site. Protestors held signs decrying what they saw as the desecration of the mountain and flew Hawaiian flags (some upright, some upside down in an early sign that the anti-TMT movement was now entangled with the Hawaiian sovereignty movement). Thirty-one protestors were arrested for blocking the road, but the size and vehemence of the protests prompted Hawaii's governor to temporarily postpone construction. When construction vehicles tried to return in June, protestors built several *ahu*—Hawaiian ceremonial shrines made of stacked stones—in the road and at the TMT site and objected to their destruction on religious grounds. (One was bulldozed, and others in the road were eventually voluntarily removed.) On the same day, a buried fiber optical cable linking the Mauna Kea observatories to a sea-level network was tampered with.

The fight against the TMT also quickly reached social media. *Game of Thrones* and *Aquaman* actor Jason Momoa, who is of Native Hawaiian descent, heard about the protests and posted a shirtless photograph of himself on Instagram with WE ARE MAUNA KEA written across his chest. Several other actors soon followed suit, and #WeAreMaunaKea became a popular hashtag. The high-profile nature of the social media conversation helped the Mauna Kea protests make the national news. Just as Mount Graham had been simplified to squirrels versus 'scopes in the media, the 2015 Mauna Kea protests were quickly dismissed as religion versus science, an oversimplification that did neither side any favors.

In the aftermath of the vehement protests and media storm, a legal case objecting to the issuance of the TMT's building permit reached the Hawaii Supreme Court in August 2015. The court ultimately invalidated the permit on the grounds that it had been issued too quickly, before an earlier hearing had been resolved. By the end of the year, construction on the TMT still had not begun.

o o o

I spoke to a number of my fellow astronomers about the TMT in 2018 and early 2019, around four years after the dramatic and contentious protests of 2015. It was, interestingly, the only subject that people regularly refused to speak about in interviews or on the record, highlighting how fraught and volatile the topic had become.

A few astronomers complained that the protestors simply couldn't be reasoned with at this point. They expressed frustration at those who were anti-TMT even after years of legal decisions and concessions on the part of the TMT. There also seemed to be persistent untruths popping up again and again about the effect the telescope would have on Mauna Kea's ecosystem. The telescope, it was claimed, would drill seven stories into the mountain and poison the water table, statements that the TMT and environmental impact studies had debunked. Other opponents described the TMT as a massive, sprawling five-acre building, when in fact, the observatory's five-acre footprint encompassed the dome, support building, gravel parking lot, and any area temporarily disturbed during construction. One oft-repeated online rumor claimed the telescope would be nuclear-powered. Unable to ever fully quash these statements, some astronomers thought the best approach was to throw up their hands and wait for a legal victory.

Thayne Currie saw it differently. An astronomer working on Mauna Kea and the Big Island in 2015, he was a staunch supporter of building the TMT but also adamant that astronomers should advocate for and justify their plans to anyone who would listen. At the height of the protests, he spoke to the participants and found that many of them were happy to talk to an astronomer. Most asked questions about the TMT and were receptive to Thayne's answers, debunking claims about poisoning a nearby aquifer and explaining why a thirty-meter mirror would be such a powerful addition to Hawaiian astronomy. His sense was that many of the protestors' actions stemmed from a sincere desire to protect Mauna Kea, and he earnestly believed that communication and compromise, with both sides willing to listen to each other, was the only acceptable way forward.

Thayne and some colleagues ran an informational booth at the Hilo Farmers Market for a year. Unaffiliated with the TMT in any official capacity, they printed flyers with their own money, staffed the booth weekly beginning at 7:00 a.m., and ignored colleagues who fretted that it might be unsafe for them to come face-to-face with angry protestors. On the contrary, most of the people they spoke to in the community were eager to get accurate information about the telescopes and happy to speak directly to an astronomer about the issue. Many of the people he spoke with supported or came to support the TMT. Others didn't, but the conversation continued anyway.

Today, Thayne remains convinced that dialogue is the only way forward, and he continues to urge pro- and anti-TMT people alike to listen to the voices closest to the conflict: the residents of the Big Island and the people who are familiar with the facts of the TMT and what it will bring to the community. He notes that while some protestors are staunch hardliners for whom the only acceptable outcome is no TMT (or no telescopes on Mauna Kea at all), many are willing to seek a compromise and are worth listening to.

He saves his harshest criticisms not for the protestors he has conversed with but for astronomers who have chosen to side against the TMT despite having little or no connection to the Big Island and the community. "I feel like they are either appropriating an issue or just giant hypocrites," he explained, referring to astronomers who had lambasted Native Hawaiian supporters of the TMT or insisted that colleagues who supported the telescope were racist.[26] In a conversation that's already fraught and complex, he saw uninformed voices jumping on a bandwagon to make a social point as fundamentally detrimental to communication.

There are, indeed, astronomers who are anti-TMT. One of them is John Johnson, who spent time at the University of Hawaii as a postdoctoral researcher and has previously observed exoplanets with the Keck telescopes on Mauna Kea. I asked him about the roots of his opposition to the TMT, and he cited the racial tensions and Hawaiian history present in the debate. John saw the protests as, first and foremost, a conflict over sacred land that had been seized from indigenous Hawaiians, an indignity that was no less potent with the passage of time. He dismissed criticisms that his argument was negated by his prior use of other telescopes on Mauna Kea for his own research. As he saw it, he was morally obligated to support the protestors' right to halt the telescope—one of the last vestiges of resistance available to them—rather than staying silent and being complicit in ongoing oppression. "There's nothing that can be discovered with the TMT that can make this worth it."[27]

John and Thayne agree on only one single point, which is this: these protests are not, in the end, about a telescope.

o o o

By mid-2019, the TMT had emerged from four more years of legal battles, culminating in a second appearance before the Hawaii Supreme Court and a 345-page report that had ultimately reissued a new permit with more conditions attached near the end of 2018.

On top of their previous agreement to remove three existing telescopes on Mauna Kea before the TMT began operations, the University of Hawaii now promised to remove two more, including one of two operational telescopes that had recently participated in a worldwide effort to take the first direct picture of a black hole. The TMT agreed to adopt a zero waste management policy, trucking all waste products off the mountain to further minimize their environmental impact. All TMT employees would receive mandatory cultural and natural resources training, and the TMT would contribute $1 million annually to a community benefits package to support projects on the Big Island. Thanks to the efforts of Thayne and other astronomers on the Big Island, public support for the TMT had increased. After the final Hawaii Supreme Court decision, Thayne continued to advocate for ongoing communication with telescope opponents. A new permit was issued, and construction was scheduled to begin on July 15, 2019.

That day, hundreds of protestors once again blocked the Mauna Kea Access Road. The 2019 protests, while still citing religious rights and environmental concerns, now heavily featured upside-down Hawaiian flags along with chants and signs declaring that Hawaii was not an American state but a sovereign nation that had been illegally occupied by the United States for more than a century. The protests criticized the TMT as a symbol of colonialism and white supremacy and sought to halt its construction as an act of Native Hawaiian self-determinism. On the third day of the protests, thirty-three Native Hawaiian elders were briefly arrested for blocking the road, but as they were led away one by one, a larger line of women blocked the road in their place with no further arrests made. The roadblock and protest would remain in place for months.

Faced with limited and unreliable road access, all the telescopes on the mountain evacuated their staff shortly after the road was blocked. No observing was done for four weeks—the longest Mauna Kea Observatory had ever sat idle—before a limited access policy was eventually negotiated with the protestors. The futures of the TMT and the mountain remain uncertain.

○ ○ ○

In all the protests against the TMT, there have been no signs or chants or hashtags aimed at astronomy itself. The protestors are not decrying telescopes themselves as forces of evil. They are not lambasting some perceived horrors of astronomical research or insulting scientists because of what they study.

What the protests *are* about changes depending on whom you ask: environmental protection, cultural rights, religion, sovereignty, or simply being able to exert power in a fight for the mountain. For many protestors, it's a mixture of all of these, but it's never been about the actual TMT itself. The question now seems to be whether the TMT and other telescopes like it are doomed collateral damage in these debates or whether they can come to coexist with their detractors, ultimately being built but in a way that respects the mountains they're on.

It was very rare, in writing this book, for me to interview an astronomer who didn't at some point speak romantically, almost reverently, about the sheer beauty and rarity of the mountains where we work. We all visit observatories to do a job, one that can involve hours spent with zeroes and ones and grappling with fussy computers and ornery mechanical parts, but none of us are blind to the raw beauty of the hillsides and sunsets and spectacular skies surrounding us when we observe. I never once heard anyone take for granted how rare and special these places are.

In a community that so deeply values the planet we're on, the summits we visit, and the human curiosity we bring to the skies, I have to hope we can find a way to respect and share our own humanity, our knowledge of the cosmos, and our love for the mountains that make our work possible. They're the windows we're able to climb to that give us a glimpse of the universe.

CHAPTER SEVEN

HAYRIDES AND HURRICANES

During the summer after my junior year of college, I worked as a tour guide at the Very Large Array (VLA) radio astronomy observatory in New Mexico. I'd read about the place and seen pictures of it for years before ever actually setting eyes on it myself, and on my first day, I was bursting with excitement. I'd driven out from Socorro, where I had just moved in to spend my summer as a research student at the National Radio Astronomy Observatory (NRAO) operations center, and dutifully geared myself up to lead tours through grounds of the observatory. I was wearing steel-toed boots, sensible jeans, and an NRAO T-shirt, and I had been issued a two-way radio and a lightning monitor. (The latter would warn me if there were lightning strikes nearby on the high plains so that I could bring my tour back to the safety of the visitor center.) I clipped the radio and monitor to my belt, neatly tucked my T-shirt into my jeans, and reveled in my sheer scientist-level dorkiness. I was so excited, I could barely stand it.

This first tour was very early in my time at NRAO, and I'd come armed with a broad array of facts about the VLA. It had twenty-seven antennae (plus a twenty-eighth to serve as a spare that could be brought in if another dish was being worked on), each a radio telescope in its own right, arrayed in a giant Y across the volcanic plains of the San Agustin Basin. Each bright white antenna was (check notes in pocket) ninety-four feet tall and weighed 230 tons; the eighty-two-foot-diameter aluminum dish of each antenna could have fit a baseball diamond inside. Each antenna was also designed to be moved along a set of railroad tracks to change the size of the Y and the configuration of the telescopes. The array had A, B, C, and D configurations; the widest, A, gave it arms that were each thirteen miles long, while the most compact, D, had been nicknamed the "*Contact* configuration" since, as the tightest and most photogenic, it was the setup captured by camera crews for the 1997 film. The different arrangements were key because the radio data captured by each of the antennae would ultimately be combined in a central operations building, a process known as (check notes again) interferometry, so they could act as a single telescope with a diameter that matched the array's total size. In the A configuration, the array could simulate a single telescope that was twenty-two miles across.

I tucked my notes back into my pocket. I was ready! The VLA, remote as it was, still got a regular trickle of tourists, ranging from amateur astronomers and ham radio enthusiasts to a quirky overflow of people who were

The Very Large Array in New Mexico (in its photogenic "D" configuration). *Credit: Alex Savello, NRAO/AUI/NSF.*

already in central New Mexico to visit Roswell, that famous wellspring of alien conspiracy theories. As tour guides, we were mostly asked about the standard facts of the telescopes and the objects they studied in space. I was prepared for those questions.

First, though, I had a few of my own. As part of my summer research program, my fellow students and I would get the opportunity to propose for observing time on the VLA. A few hours of telescope time had been specifically set aside for our use, and I wanted to use it to study a few red supergiants. I knew that these stars could shed mass from their outer layers and that this mass would hang around, enshrouding the star in a sort of dusty shell as it cooled and dissipated. Apparently, the dust shell could sometimes produce something called a maser, stimulating and amplifying the emission of radio light at very bright and very specific wavelengths from molecules in the dust, but the detailed physics behind how this worked was still a mystery. I was curious as to whether we could explain how the masers formed and whether this could tell us anything about how the stars were dying, and with access to a radio telescope, I finally had the chance to observe these stars myself. We had the whole summer ahead of us to learn about how radio telescopes worked, but I wanted to get a head start. Plus, I'd used optical telescopes before thanks to my work with Phil Massey, and I'd taken Jim Elliot's observing class. How different could radio telescopes be, really?

One of the on-site astronomers had agreed to chat with me, and I started out by asking a few questions about what they were currently observing, figuring that'd be a good start.

"What are you observing?"

"Well, right now, we're on a phase calibrator for water near a protostar."

Hmm. I tilted my head a little bit and nodded like this made sense.

"Mmm...how bright is it?"

"About four Kelvin."

I...what? Kelvin was a unit of *temperature*, not how bright something was. My head tipped a little more.

"Er...how's the observing going?"

"Well, the receiver is centered at minus thirty kilometers per second, but if the protostar is less than a millijansky per beam..."

This was starting to feel like an Abbott and Costello skit; I was fairly sure that I'd heard all these words before but had no idea they could be put together in this order. The angle of my head had listed past "thoughtful head tilt" and was rapidly approaching "confused schnauzer." My comprehension level wasn't faring any better.

The observer happily continued to chat about his science. We were both astronomers, but the terminology whizzing around felt like I'd stepped into a different field entirely. I hung on as best I could but felt a little bit of dread settling in my stomach. How different *was* this research, just because we were working at longer wavelengths? Very, apparently. (Happily, by the end of the summer, after ten weeks of research and a radio-astronomy-for-schnauzers crash course, I'd be chattering away with the rest of them.)

"We're actually just about to switch targets and go to..." The observer kept talking, but this statement stuck in my mind, and something clunked into place: the array was observing *right now*. I'd known this, intellectually, but it hadn't really settled in my bones that I'd been standing among antennae that were currently capturing data, collecting radio light from stars and galaxies that were invisible behind the sun's glare to my optical-light-detecting eyes but very much observable for a radio telescope, where the sun is relatively dim. As I headed back out from the building to pick up my first tour group, I filed this away as a fact I should pass on to them. We weren't just standing among scientific instruments. We were standing in the middle of science that was *actively happening*. Such is the weird and fantastic world of radio astronomy.

o o o

Radio telescopes like the VLA or the ill-fated 300 Foot telescope in Green Bank, West Virginia, don't, at first glance, look much like "normal" telescopes, the shiny, mirrored setups ensconced in domes and carefully brought out for observations only at night. Most of them don't have domes, and while there's at least a curved surface that somewhat resembles a telescope mirror, they look less like shiny reflective surfaces and more like giant metal bowls.

The trick, of course, is that we're looking at them with human eyes. Radio telescopes work at the far end of the electromagnetic spectrum, capturing light with wavelengths measured in millimeters, centimeters, even full-on meters, wavelengths far past the narrow regime that our eyes can detect. For wavelengths this long, the surfaces of these radio telescopes *are* shiny; radio light pouring down from the sky will bounce off a radio dish and get focused into a detector in much the same way that optical light does on a traditional mirror.

At such long wavelengths, it's also easier to use a technique known as interferometry, the process that lets the many antennae of the VLA function as a single telescope. In interferometry, an array of telescopes in different locations—separated by as little as a few feet or as much as an entire continent—will act like shiny spots across a giant virtual mirror. Astronomers can point these telescopes at the same object, record radio data from that object at each telescope, and then send the data to a single facility that can computationally construct a final image. It's a complex technique, but combining the data is easier at longer wavelengths, making radio astronomy ideally suited for this approach.

The payoff is also worth it. Larger distances between the telescopes give us a wider virtual mirror and a sharper final image, even though the "mirror" is mostly made up of empty space (this is why the VLA antennae are spread out along railroad tracks and can be arranged into four different configurations). In 2017, radio telescopes across the world achieved the ultimate interferometric victory, combining data from telescopes in Arizona, Hawaii,

Mexico, Chile, Spain, and the South Pole to produce a planet-sized telescope and take the first picture of a black hole. The result—an image of a black hole 6.5 billion times as massive as our sun in the heart of a galaxy fifty-three million light years away—made headlines across the world when it was released in April 2019.

Even with just one single radio telescope, we open up a whole host of new science opportunities by moving into this wacky but powerful wavelength regime. Observing radio light lets us study everything from Jupiter's magnetic field to the birthplaces of new stars and even the fading background radiation left over from the Big Bang.

We also get some incredibly cool-looking observatories. The VLA played a starring role in the movie *Contact* and has appeared in everything from music videos to telecom commercials. Another radio telescope featured in *Contact* is Arecibo Observatory, an immense thousand-foot dish built into the natural depression of a sinkhole in northwestern Puerto Rico. While this design means the dish itself is immobile, it's still able to work as a fully functioning telescope, pointing at its scientific targets by moving the receiver suspended nearly five hundred feet above the dish on a trio of enormous cables to observe objects passing overhead. James Bond fans are also likely familiar with Arecibo, though they may not realize it: in *GoldenEye*, the observatory played the secret antenna concealed under a lake and used by the film's villain to control the titular satellites. At the film's climax, Bond (played by Pierce Brosnan) and the villain (played by Sean Bean) fight on the precarious receiver platform until Bond sends his opponent plummeting down to the dish (adding Arecibo to the illustrious but lengthy list of "things that have killed Sean Bean in movies").

o o o

The long wavelengths we're working with in radio astronomy mean that radio telescopes are capable of all sorts of observations and tricks that normal telescopes can't do, beyond the already impressive feats of interfero-metric cooperation and imaging black holes.

As I freshly realized during my first day as a tour guide at the VLA, radio telescopes can quite effectively observe during the day. Radio obser-vations can happen through sunshine, clouds, and rain, and even wind only becomes an issue at its most vicious if it shakes an antenna hard enough to physically mess with signal quality.

Even snow isn't a problem in moderation; plenty of radio telescopes happily keep observing through snowfall and even blizzards. The only real difficulty comes when snow starts to accumulate in the dishes, since the weight can bow the dish or put strain on the motors. Different radio observatories have different approaches to clearing off snow. Green Bank briefly investigated building a fire under the telescope to serve as a source of heat. (It was quickly extinguished—literally—when drips from the freshly melted snow put out the fire. Astrophysicists, ladies and gentlemen.) Green Bank even took a crack at blasting the 300 Foot telescope with a Rolls-Royce jet engine. The VLA has a "snowdump" command for controllers that simultaneously tilts all twenty-seven dishes down to their lowest limits to physically dump the snow out of the bowl, spins the dishes into the wind to blow the snow out, and points them at the sun to melt any ice. Other radio telescopes take the considerably lower-tech approach of what several colleagues described as the "grad student with a broom" method.

If someone took a broom or a jet engine to an optical telescope's mirror, the observatory engineers would throttle them, but these approaches are completely acceptable for radio telescopes. Remember, a telescope's reflec-tive surface should be shaped to a precision within 5 percent of the wave-length of light that it's focusing. This means that while mirrors reflecting short-wavelength optical light must be within a fraction of a hairsbreadth of perfection, telescopes working at radio wavelengths have a margin of error as large as a few millimeters or so. It doesn't sound like much, but it means that radio telescopes, under supervision, can actually be *walked on*. A highlight of my summer at the VLA was being taught how to safely walk on one of the antennae (keeping our feet on the seams and officially taking care to not climb too close to the lip of the dish), and I got to take my parents and Phil Massey and his family up into an antenna for a tour. At the Parkes Observatory radio telescope in Australia, astronomers can sit on the edge of the antenna and take a high-flying ride as it's raised into the air, a trip known as a "hayride."

During Hurricane Maria, Arecibo was actually able to observe during parts of the devastating storm, helped along by astronomers and staff who had hunkered down at the observatory to ride out the 155-mile-per-hour winds. Afterward, the observatory site served as a relief center once roads were cleared, providing water and basic services to local residents and serving as a staging area for FEMA efforts thanks to its on-site helicopter landing pad. Unfortunately, the telescope didn't come away entirely unscathed. The telescope's line feed—a ninety-six-foot-long radio receiver, with the appearance of a long tube-shaped ladder—was torn off by the raging hurricane winds and fell through the mesh dish of the telescope, punching several holes. The Arecibo dish itself is actually suspended above the ground, high enough that it's possible to walk and drive under the dish. The mesh that looks so solid from a distance or in aerial shots actually lets light filter through, enough for a thick, green carpet of plants to grow under the telescope. The area underneath flooded after the hurricane, and crew members wound up borrowing an astronomer's kayak to inspect the dish from below.

Still, radio telescopes share some of the same problems that plague any other observatory; they just manifest in slightly different ways. Animals can get up to just as much mischief at radio telescopes as they can anywhere else but with a few variations on what can cause serious problems.

The most famous pests at radio telescopes are birds, or rather what birds can leave behind. In 1964, physicists Arno Penzias and Robert Wilson were working on an extremely sensitive radio antenna and grappling with how to get rid of a persistent background hiss in their data. One of the early culprits

they considered were the pigeons that had been nesting in and near telescopes. The birds were producing a coat of what Arno politely referred to
as a "white dielectric substance," a problem for radio telescopes in particular
since bird droppings can transmit electrical signals and mess with detectors. Arno and Robert cleared out pigeons and poop alike in the hopes of
fixing the background noise. The bad news was that the hiss remained; the
good news was that this proved to be the cosmic microwave background,
an electromagnetic relic of the Big Bang and a discovery that netted the two
physicists a Nobel Prize.

Still, keeping flocks of birds away from radio telescopes has since
become a priority, and observatories use everything from metal spikes to
(radio-transparent) GORE-TEX shields to sound in their efforts to repel
birds and keep their droppings off the telescopes. In an ironic coinci-
dence, Jodrell Bank, a radio observatory in England, has become great at
this thanks to a pair of wild peregrine falcons that have set up house in the
support towers of their largest telescope and do an excellent job of keeping
smaller species at bay.

Arecibo fights back against birds a bit more dramatically, albeit unin-
tentionally. The telescope's transmitters, receivers, and other optical elements are housed inside a large dome suspended over the enormous main
dish with an open base. Birds have a tendency to fly in and get confused in
the dome, which doesn't always end well for them. If the birds are between
the telescope's optical elements at the wrong time, they essentially get flash-
microwaved. Most radio telescopes don't have this problem since the optical elements aren't contained in a large dome, but cases of bad timing do
occasionally happen: one radio observatory in California happened to turn
on a transmitter at precisely the wrong time and wound up zapping a passing swarm of bees.

While studying radio-bright regions of a nearby galaxy, Norbert Bartel
and his team temporarily lost data from one California radio telescope. In
their published research paper, Norbert dryly explained that the loss of signal

was "due to the presence of a red racer snake (*Coluber constrictor*) draped across the high-tension wires (33,000 V) serving the station. However…we do not consider [the snake] responsible for the loss of data. Rather we blame the incompetence of a red-tailed hawk (*Buteo borealis*) who had apparently built a defective nest that fell off the top of the nearby transmission tower, casting her nestlings to the ground, along with their entire food reserve consisting of a pack rat, a kangaroo rat, and several snakes, with the exception of the above-mentioned snake who had a somewhat higher density."[28] In the realm of unusual observing reports, "a bird accidentally dumped a snake onto the power lines" has to be one of the strangest.

Arecibo also has a high population of feral cats, and they understandably gravitate to a site where people are willing to feed them. When the cat population started to get out of control, a few astronomers started taking in kittens and quietly dispersing them to colleagues on the mainland. The network of astronomer cat traffickers spiraled and extended to cats found at other nearby radio observatories; at an astronomy building in Tucson, while investigating what sounded like rats, someone pushed up a ceiling tile and had two kittens fall out of the ceiling. They were quickly named Phobos and Deimos after the two moons of Mars. Today, the kittens of astronomy have an online presence set up at @ObservatoryCats on Twitter; the account has raised funds to help care for the pet population of Puerto Rico after Hurricane Maria and keeps track of astronomy cats that got their start at telescopes.

o o o

Even with the occasional pooping bird or fried cloud of bees, it's tempting to think of life at a radio observatory as pretty cushy. You can observe through any weather, climb around on the dishes like giant jungle gyms, and apparently, it even rains kittens.

The catch comes when we consider noise. We know that radio

telescopes have different requirements than we might first imagine for what constitutes a "shiny" mirror; it turns out they have similarly odd criteria for what we'd consider "dark." If you could see radio light, almost anywhere you're sitting while you read this would be a chaotic blend of strange signals. Sitting in a city coffee shop as I write this paragraph, if I could pick up various radio wavelengths with my eyes, I'd be seeing a cloud of Wi-Fi networks, endless blasts of signal from cell phones, the occasional flash of the shop's microwave, and even blinks of light from spark plugs in the internal combustion engines of cars driving by outside.

Radio telescopes need to shield themselves from these contaminating light sources as much as possible. Entire spans of wavelengths are actually protected via national and international regulations, restricted so they can't be used by anyone (broadcasting radio stations, military communications, and so on), to leave them pristine and undisturbed for scientific research. Still, the more remote a radio telescope is the better. The plain the VLA sits on is surrounded on almost all sides by mountains, protecting it from encroaching radio sources that could drift in from nearby population centers. Plenty of observatory summits that host optical telescopes will also have a radio telescope or two on-site, simply because it's already a convenient and well-developed remote site. Kitt Peak has a twelve-meter radio telescope and one antenna of the Very Long Baseline Array (a continent-sized radio interferometer with ten antennae spread across the United States), Mount Graham hosts the Submillimeter Telescope, and Mauna Kea Observatory features a fifteen-meter submillimeter-wavelength radio telescope, another Very Long Baseline Array antenna, and the Submillimeter Array, an eight-antenna interferometer.

Still, some places push remoteness to an even further extreme. Recall the National Radio Quiet Zone in West Virginia, where Green Bank Observatory (and its ill-fated 300 Foot telescope) is located; in the area closest to the telescope, Wi-Fi, cell phones, and microwave ovens are all banned, and all vehicles run on diesel engines. However, the payoff is significant: the site is so excellent for radio astronomy that after the 300 Foot telescope collapse, Senator Robert C. Byrd of West Virginia advocated for its replacement and pushed funding through Congress to build a new enormous radio telescope in Green Bank. The 100 Meter Robert C. Byrd Green Bank Telescope is currently the world's largest steerable telescope and is still observing today.

o o o

Preventing people from even using cell phones in the presence of radio telescopes may seem extreme, but radio astronomy can be particularly vulnerable to spurious signals, and separating the commonplace from the cosmic in these telescopes' data is no easy task.

The 210-foot Parkes Radio Telescope sits in a rural sheep paddock about 225 miles west of Sydney, Australia. The largest telescope in the Southern Hemisphere, it briefly became famous when it received and broadcast footage of the 1969 Apollo 11 lunar landing to the world, but it has also mapped the hydrogen gas of the Milky Way and discovered thousands of new galaxies.

For years, Parkes also detected funny flashes known as perytons. Named for a mythological creature with a winged stag body that cast the shadow of a human—a creature that looked like one thing but turned out to be something else—the perytons appeared in the Parkes telescope data as brief blasts of radio light. Over the years, they'd been spotted pretty much all over the place and had been reported only during weekday office hours. Since the universe doesn't generally care about business hours, it was widely accepted that they were coming from some sort of nearby noise source on the ground, but nobody had sussed out what.

Perytons were a known oddity and nuisance at Parkes when Emily Petroff began using the telescope for her research as a graduate student in 2012. The trouble, in Emily's case, was that she was studying brief blasts of radio light that really *were* originating from deep space, strange signals known as fast radio bursts. These bright radio flashes seemed to be coming from mysterious and unexplained astrophysical phenomena, but when Emily was studying them, they were also shrouded in skepticism: how did we know that fast radio bursts didn't fall into the same category as the unexplained but clearly earthbound perytons?

It was a fair question to ask and one that had been posed before in radio astronomy. Jocelyn Bell Burnell had faced a similar concern when she was a graduate student in 1967, following her discovery of a mysterious and fascinating signal in data from a radio telescope in Cambridge, England. Pulses of radio emission had appeared in her data, arriving just over once a second. The pulses appeared with astonishing regularity, like a perfectly ticking clock, and were unlike anything astronomers had ever seen from the sky. The signal produced such a precise series of pulses that at first, Jocelyn and her colleagues jokingly named the first four pulse sources LGM-1 through LGM-4. The acronym stood for "little green men."

Jocelyn was wholly aware that interference from the ground could be mistaken for something like this and kept a careful eye on how the sources moved over the course of the night. In her case, she quickly realized that the pulses she had discovered were bona fide astronomical sources; she observed her first signal for months and discovered that it rose in the evenings along with the rest of the night sky, tracking the apparent motion of the heavens as the earth turned. Jocelyn had discovered objects known as pulsars, the rapidly spinning leftover cores of dead stars that were shining bright beams of radio light along their magnetic poles and acting like cosmic lighthouses. The slowest pulsars emit a few radio pulses per minute, while the fastest emit hundreds of pulses in a single second, spinning faster than a hummingbird beats its wings. The discovery was recognized with a Nobel Prize (although once again, the committee avoided giving the prize to a woman, instead recognizing her thesis adviser and another colleague), and Jocelyn went on to a long and distinguished research career, receiving a

2018 Special Breakthrough Prize in Fundamental Physics to recognize her scientific achievements.

Back at Parkes in 2014, Emily Petroff was eager to get to work on fast radio bursts, the newest puzzle of radio astronomy, and decided that her first challenge would be to solve the peryton mystery. She developed a technique for quickly finding perytons in the radio telescope's data and quickly collected dozens over a two-month period. The timing of this larger sample gave her and her team their first clue: the perytons were almost always detected around lunchtime.

The entire Parkes Observatory site staff got in on the action. The team would position the telescope at a spot where it had previously detected a peryton, and the staff would dash off to the neighboring administrative buildings and do whatever they could to generate a peryton: opening and shutting doors, testing magnetic locks, plugging and unplugging computers, and running just about every bit of equipment they could think of. Eventually, Emily's team was detecting perytons almost in real time and calling the staff to ask what they'd been up to: Had there been a camera crew on-site? Was construction work happening nearby? The staff kept experimenting, and the flashes still appeared randomly now and then, but nothing they tried was able to make a peryton on command.

The next breakthrough came when the staff checked a recently installed radio interference monitor and noticed that the perytons were happening alongside a blast of radio light typically associated with electronics. The telescope hadn't detected it and for good reason: there wasn't any scientific point in observing wavelengths that were already dominated by electronic signals. After studying this new data and recognizing the electronic signal, the mounting evidence Emily's team had amassed on perytons now pointed to two clear culprits: the two microwaves in the Parkes Observatory kitchens.

The staff sprang into action again, pointing the telescope accordingly and then running both microwaves every which way they could. They operated the microwaves for a few seconds or a few minutes, microwaving nothing or a mug of water or someone's lunch, but they still couldn't produce a peryton on purpose. Finally, someone hit upon one last idea. What if they used the microwaves less like careful scientists and more like hungry staff members waiting for their lunch? What if, instead of letting the microwave finish its cooking cycle during every test, they stopped the microwave like so many thousands of other impatient users did, opening the door to stop the microwave while it was still counting down its final few seconds?

The test worked. Three hastily opened microwaves generated three clear blasts of perytons. Emily was flying back to Australia from a job interview when the test results came in, got the news via email at the Singapore Airport, and wrote the entire scientific paper summarizing the discovery in an elated four-hour burst on her final flight home. The Parkes Observatory site staff all earned coauthorship on the paper.

Today, perytons are a known quantity, and real fast radio bursts are still being discovered and studied by Emily and other experts in the field. We know now that they're highly energetic, that they're coming from brief and dramatic astronomical events in other galaxies, and that they're not microwaves, but what exactly *is* causing them remains a mystery.

o o o

Parkes picked up microwaves. Other radio telescopes risk detecting Wi-Fi or cell phones. Green Bank avoids spurious signals by isolating itself in a radio-quiet corner of the globe. Still, the challenges radio telescopes face are just another manifestation of a bigger problem: the earth is a tough place for astronomy. Whether it's human-made signals messing with radio antennae, light pollution brightening dark skies around observatories, or water vapor and a turbulent atmosphere interfering with the light traveling to our telescopes from the heavens, working on the surface of the earth comes with a host of challenges. It's undeniably tempting to imagine sending all our telescopes up to join Hubble in space, but while this sounds appealing, it's both financially and logistically implausible. Ground-based astronomy remains an economical, nimble, and accessible option for astronomers, particularly when we start to get creative…or stretch the definition of what exactly we can call "ground-based."

CHAPTER EIGHT

FLYING WITH THE STRATONAUTS

The unicorn was displeased.

The status unicorn for the SOFIA telescope sat perched on the instrument station. It was one of those intensely adorable little round stuffed whatsits that could be turned inside-out to display two different faces and was "operated" by the telescope's instrument scientist. When things were going well, it displayed shiny white fur and a perky smile. When things weren't going well—like tonight—the unicorn was turned blue-fur-side out, displaying a disappointed frown and big disgruntled eyes.

Ad hoc or not, it was a good gauge of how things were looking for the telescope I was visiting. Nothing was definitive yet, but it sure looked like we were about to lose our planned night of observations due to either equipment problems or weather. A cooling unit was acting up, and if a fix wasn't found, it would render the telescope incapable of observing. It was also atypically cold outside: it was February 2019, and even being in southern California hadn't kept the temperature from plunging close to freezing. If any part of the observatory started to ice up, that would be it for the night.

The group of people milling around and waiting for official word to

The Stratospheric Observatory for Infrared Astronomy (SOFIA) in flight with the telescope chamber open. *Credit: NASA/Jim Ross.*

come back before the night started was a bit antsy but mainly adopted the cheerfully fatalistic attitude of observers everywhere: there wasn't a whole lot any of us could do but wait. People chatted in groups, checked the forecast on their phones, wandered over to the instrument panel to check on the unicorn, and broke into the night lunch supplies that were nominally supposed to be saved for later. Eventually, cutting through the thick air of waiting, the word came back.

No flight tonight.

A problem with cooling systems, encroaching weather, and the early consumption of night lunch may have all been ordinary observing scenarios, but the telescope we were currently waiting in was anything but. About twenty of us were onboard the Stratospheric Observatory for Infrared Astronomy—SOFIA—a modified Boeing 747-SP with a 2.7-meter telescope mounted in the back. When everything was going well, the plane was designed to fly into the stratosphere, as high as forty-five thousand

On board SOFIA during my first visit to the observatory. I'm on the right in a hooded jacket; the sealed-off telescope chamber can be seen in the upper left. *Credit: David Pitman.*

feet, and then raise a 13.5-foot-wide retractable door on the rear left side of the plane to expose the telescope. Operating above 99 percent of the water vapor in the atmosphere, it would observe a span of wavelengths that would normally bounce off water molecules and be impossible to capture from the ground. The rest of us—pilots, mission directors, telescope and instrument operators, safety techs, and observers like me—rode along with the telescope, operating it from the pressurized part of the 747 while the telescope sat just next to us, sealed off in its own open-doored chamber.

The critical cooling system failure was coming from inside the telescope chamber. SOFIA is able to observe, rock-steady, through turbulence because it's mounted on the world's biggest ball bearing, 1.2 meters in diameter and sealed into the mount of the telescope along with a generous supply of oil that lets it float smoothly. Without the coolant to keep the oil from heating up under the friction of the moving bearing, the telescope would eventually stop floating on the bearing and be subject to the same bumps and rattles of any normal plane flight, jarring it and its field of view to the point where it would be rendered unusable.

On another night, the crew might have opted to work on the cooling for a little while longer with SOFIA on the ground, but we were also racing the weather. Despite being based in Palmdale, California, just north of Los Angeles, we'd been hit with a freak storm and cold snap, and the wings of the plane were at risk of icing up. While normal commercial planes can simply deice their wings, that wasn't an option for SOFIA. Anyone who's been in a freshly deiced plane has seen the viscous neon-colored liquid that gets sprayed over the plane and the droplets that continue to drift backward during the first few hours of the flight. In SOFIA's case, deicing would risk some of the fluid flowing backward during the flight and getting into the telescope chamber once it was opened. For us, this meant that any ice forming on the wings would immediately ground the telescope.

The bizarre weather had also grounded SOFIA the previous evening; the first observing flight I'd been scheduled on had been canceled due to a risk of extreme turbulence. Incredibly, despite combining all the complications of a large aircraft *and* a large telescope, these sorts of weather and equipment cancellations are actually quite rare; two in a row was (literally) a unicorn event. It was simply pure bad luck that it had happened on the two flights I'd been scheduled for as a passenger. As I filed sadly off the plane and back into the nearby hangar, I worried I might have become one of those unlucky astronomers with a weather curse, dooming SOFIA to remain on the ground with my mere presence. I also suspected I'd just lost my only chance to fly on a telescope.

o o o

In recent years, one key challenge for ground-based telescopes has been overcoming the irksome presence of our planet's atmosphere.

The twinkling that makes stars so pretty for many of us stargazing on the ground is a perpetual problem for astronomers—the source of the seeing described earlier—and attempts to minimize it have gotten truly spectacular thanks to the advent of adaptive optics. An adaptive optics system places a computerized system of magnets behind shaved-thin telescope mirrors and fires a laser into the upper atmosphere. The laser excites atmospheric sodium atoms, making them glow and producing an effective fake star. By comparing the fake star's appearance to the theoretical perfect image they should get from the laser, the system can measure the atmosphere's distortions and use the magnet system to tweak the shape of the mirror and compensate for the atmospheric effects in real time. The result is a gorgeously sharp image that mimics how the sky would look without the atmosphere in the way.

It's a brilliant method and allows the sharpness of images taken at telescopes with adaptive optics systems to surpass those taken with space telescopes like Hubble. Still, this solves one inconvenience visited upon us by the air we breathe but not another.

Our atmosphere, in addition to stirring up the light that actually makes it to the earth's surface, also blocks vast swaths of light from ever reaching us in the first place. Most light at shorter or longer wavelengths heading our way from outer space—invisible to our eyes but invaluable if we can capture it with a telescope—finds itself blocked by the atmosphere on its way here. Exactly what does the blocking varies by wavelength. Gamma rays and X-rays, with their immensely high energies and tiny wavelengths, ping off our upper atmosphere. Some ultraviolet light leaks through in small amounts (enough, at least, to give us all the occasional sunburn) but is mostly stymied by the oxygen molecules we know as the ozone layer. Similarly, some infrared light can make it through, but longer infrared wavelengths are blocked by atmospheric molecules like water vapor and carbon dioxide. It's not until we get well into the radio regime—submillimeter wavelengths and longer—that light starts leaking through the atmosphere again.

Studying light at the edges of these limits, particularly around the infrared and submillimeter wavelengths, is possible from Earth but requires getting ourselves away from as much water vapor as possible. Even the gains afforded to us by high altitudes or dry observatory sites can be enough to vastly improve matters for telescopes trying to observe these types of light.

This is one reason why Chile is such a spectacular place for astronomy. The Atacama Desert is famously the driest place in the world apart from the poles. The optical observatories in Chile are largely situated in the central and northern foothills of the Andes, but in the high planes of the Atacama, we've built observatories like the Atacama Large Millimeter Array (ALMA), a radio interferometer consisting of sixty-six antennae. Situated at nearly 16,600 feet above sea level, it's given us our first glimpse of planets being born around young stars—coalescing out of thin disks of gas and dust that will eventually form planetary neighborhoods similar to our own solar system—and galaxies crashing into one another thirteen billion light-years away, so distant that the light detected by the antennae was emitted by the two galaxies in the earliest eon of our universe, less than a billion years after the Big Bang. The observatory also served as one element of the planet-wide interferometer that took the first picture of a black hole.

However, activity at the telescope site itself can be extremely arduous. Technical staff working on the antennae or support infrastructure are required to use portable supplementary oxygen, and the on-site observatory building has oxygen piped in. Employees sleep at the Operations Support Facility, which has a relatively low altitude of about 9,500 feet. Still, impressive as it is, ALMA's altitude is an excellent improvement over sea level but still barely nudges into the underside of our planet's atmosphere. To really see gains, we need to get higher. Much higher.

o o o

The idea of tossing a telescope in a plane and simply *flying* up to higher altitudes originated back in the 1960s. One early airborne observatory was a twelve-inch telescope mounted in the cabin of a NASA Learjet, positioned to peer out a sealed-off round hole just in front of the plane's wing. The Learjet would fly to altitudes of fifty thousand feet with an astronomer operating the telescope inside the plane while wearing a helmet and oxygen mask, and it was used to study infrared light from other planets in our solar system, newborn stars, and black holes at the centers of other galaxies.

Larger-scale airborne astronomy had a more difficult start in the form of the Galileo Airborne Observatory. Originally built by NASA in 1965, the Galileo I was a modified Convair 990 with additional windows mounted along the top of the aircraft. It was used to observe a solar eclipse and comet flyby in 1965, and NASA had hoped that Galileo I could be used as a multifaceted airborne science facility, studying astronomy as well as other research applications like wildlife surveys, for many years to come. Unfortunately, in 1973, the Galileo I was returning to Moffett Field in California following a test flight when it collided in midair with another aircraft, a U.S. Navy P-3 Orion, as both planes were trying to land. The crash killed all eleven crew members onboard the Galileo I and left only one survivor out of the six crew members on board the Navy P-3. NASA eventually rebuilt the observatory with a second modified Convair 990, the Galileo II, which continued to observe until 1985. Unfortunately, that aircraft was destroyed as well when its two front tires blew during a takeoff roll at March Air Force Base, causing a runway overrun and catastrophic fire; amazingly, despite the immense fire, everyone on board survived.

The Kuiper Airborne Observatory, named for Gerard Kuiper, a planetary science specialist and early champion of airborne astronomy who had observed with the Learjet telescope, was NASA's first sizeable triumph in airborne astronomy. The Kuiper aircraft was a Lockheed C-141A Starlifter, designed similarly to SOFIA with a retractable door and sealed chamber that held a thirty-six-inch telescope designed for long-wavelength infrared observations. The plane could reach forty-five thousand feet and flew from the mid-1970s until 1995.

The Kuiper Airborne Observatory's many discoveries include the first observational evidence of Pluto's atmosphere and the rings around Uranus. If this sounds familiar, there's a reason; in a demonstration of what a small world astronomy is, these observations were helmed by my MIT observational astronomy professor, Jim Elliot.

Jim's class was where I'd first heard tales of airborne astronomy, although I had a slightly skewed impression of what the observing would be like. When Jim described observing with a telescope from the open door of a plane, I took what I mentally knew about telescopes at the time (my dad's little backyard Celestron) and what I knew about planes (the extent of my flying experience at that point consisted of two cross-country plane trips and catching *Air Force One* on cable a few times) and just combined the two. The net result was a mental image of Jim and a few other astronomers hunched, windblown and grinning, in the open back of a plane while balancing a backyard telescope and pointing it out the door. (I guess I assumed they either held on very tightly or were tied to something.) In reality, the telescope itself, in both the Kuiper aircraft and SOFIA, is in an entirely separate chamber of the plane, sealed off from the passengers and pressurized cabin. A crew of people still flies with the telescope every time it observes, keeping an eye on the various instruments and operating it from the air-supplied side of the cabin, but nobody is actually hanging out the open side of the plane like a delighted dog riding in a car's passenger seat on the highway. I was only slightly disappointed.

SOFIA has since taken up the mantle from Kuiper as NASA's airborne observatory, upgrading to a Boeing 747-SP and a 2.7-meter telescope. SOFIA began observing in 2010; for most of the year, it flies out of Palmdale, with annual deployments to Christchurch, New Zealand, in June and July so it can observe the southern sky during that hemisphere's longer winter nights. Standard SOFIA flights are about ten hours long, and since beginning operations, it's mapped the magnetic field of the Milky Way, studied new stars being born, and searched for signs of water plumes erupting from Europa, one of Jupiter's moons.

o o o

The modifications made to the plane mean that SOFIA is officially classified as an experimental aircraft, which carries with it a whole host of extra safety rules and regulations.

My first clue that SOFIA wasn't going to be a typical flying experience or a typical observing experience came when I was sent a bundle of forms to fill out after first getting permission to fly. The forms included pages of personal information for the flight's manifest, a detailed medical clearance form, and an OSHA-mandated letter informing me that SOFIA would expose me to "hazardous noise levels." Stripped of most of the standard trappings that help to absorb sound on commercial aircraft, being inside the SOFIA plane while it's in flight can be *loud*, to the point that long-term exposure could cause hearing damage; passengers are all asked to wear earplugs, and communication on-board is done via radio headset. In addition to the paperwork, I was instructed to bring along a travel mug with a locking lid so I wouldn't risk splashing hot coffee or tea around the plane in case of turbulence; warned that although the plane would be very cold, I should avoid synthetic garments as they could pose a risk in case of fire; and told I should plan to arrive in Palmdale a day early to go through SOFIA's emergency egress training before my flight. This was clearly not like most other telescopes.

Emergency egress training itself was the usual combination of fascinating and mildly alarming that always accompanies something with the tone of "we have never had a serious incident or emergency on board our flights, but just in case, here's what you should do if the plane is simultaneously upside down and on fire." In addition to standard aircraft safety instructions—how to wear life jackets and don oxygen masks—I learned how to deploy the plane's blow-up door slides and how to use them as life rafts, and found out that the slides were equipped with "survival kits" that included a first-aid pack, an emergency beacon, a knife, and even a fishing kit and how-to-fish guide. I also got video instructions on how to escape the plane through a trapdoor in the floor at the front ("please only use this exit if the plane has come to rest upside down") or through a hatch built into the cockpit (which involved climbing out and grabbing a wire-and-handle system to rappel down the hump of the 747 to the ground). Since everyone would be free to wander around the plane during most of the flight, all passengers were also required to keep an EPOS on us at all times. Short for Emergency Passenger Oxygen System, these were little canvas packets that could be opened and unfolded into a hood with an emergency supply of oxygen in case of fire and smoke in the cabin. Instructions printed on the packet included the phrase "breathe normally." Right.

Before each SOFIA flight, everyone flying on the plane is required to attend the mission briefing, held in a conference room inside the NASA hangar shortly before boarding. Led by the flight's mission director, the briefing runs through the passenger list, the flight path and weather forecast, the current state of everything on the plane, updates on the telescope and instruments, and a brief rundown of the astronomical objects scheduled for observation that night along with a brief overview of the background science. The pilots told me later that these scientific updates were their favorite part of the briefings. SOFIA's call sign is NASA747, and after years of nighttime flights, a good number of air traffic controllers now recognize them and will, in quiet areas during low-key times of night, occasionally call up to ask what the plane is observing. The pilots are always happy to answer back "the center of the Milky Way!" or similar.

My first SOFIA flight was canceled as soon as its mission briefing started, due to a "significant meteorological event" that had been causing what was officially classified as "extreme turbulence" (turbulence that could injure passengers or damage the plane). I learned later that SOFIA had canceled another flight the previous week for the same reason. On the same night, a commercial flight flying a similar path to SOFIA's had encountered turbulence so severe, people had been sent to the hospital. Everyone may have been eager to fly, but not one of us even thought of questioning the decision.

I took advantage of the downtime after that cancellation to talk to some of the SOFIA pilots. Most of them were former commercial or test pilots, and I was curious about what it was like to fly a plane with a telescope on board and an enormous hole in the side. Each pilot swore that SOFIA flew almost exactly like a regular passenger plane. The most unusual feature of SOFIA, the telescope chamber door, is so excellently engineered that it's imperceptible when it's opened or closed during flight, without so much as a bump of turbulence. The telescope itself is heavy—seventeen tons concentrated in the rear of the plane—so the SOFIA computer racks are stored in the front of the plane along with steel plates set into the forward floor to keep the aircraft balanced.

The biggest difference the pilots mentioned between flying SOFIA and flying a regular plane was the extremely precise timing and flight path. One quirk of having a telescope recessed into the body of a plane comes when you consider how exactly to point the thing. Much like passengers peering out the windows of an airplane during a trip, SOFIA's view from inside its chamber is largely dictated by which way the plane is facing. This means that the plane's path on any given night is designed around what the telescope is planning to observe and how long it plans to observe for, which makes for some wonky-looking flight tracks tracing out zigzags or triangles or diamonds. The timing of the flight is also extremely precise: the pilots must stick to the flight's planned schedule to within just a few minutes. This means keeping close track of winds and aircraft weight and altitude and maintaining a constant and careful line of communication with air traffic controllers to safely share the sky with commercial flights. Despite this balancing act, most SOFIA flights are able to follow their planned paths perfectly, thanks in large part to meticulous planning and the skill of the pilots.

This requirement for a preapproved observing plan somewhat diminishes the job of any astronomers on the plane. The astronomers may be the people who were granted time on SOFIA and who have supplied the observing plans and targets used to design the flight's path and schedule, but by the time the plane actually takes off, they're largely just along for the ride. Observers on SOFIA are occasionally able to allocate a spare minute or two of flight time to one of their targets or twiddle an instrument setup to improve some small detail of the data, but with the flight plan already set in stone, there's very little that can be done. On the other hand, SOFIA's telescope operators perform a very similar function to their counterparts at other observatories, albeit with a few extra complications thrown in. (Most operators, even in earthquake zones, don't regularly have to deal with turbulence.)

A full SOFIA flight typically includes two pilots, a flight engineer working in the cockpit on aircraft operations, mission directors, safety techs, telescope operators, instrument scientists, and a handful of visiting astronomers; on my February flight, there were twenty people on board. Emily Bevins, one of the telescope operators, hit on a description of SOFIA that struck me as perfect after two days of tours and egress training and mission briefings, all unlike anything I'd ever been through at other observatories. "It's like a symphony," she explained, with multiple well-rehearsed groups of people each contributing their own meticulously prepared parts to create a complex but cohesive piece of music. The analogy hit home wonderfully for me as a violinist who had played with orchestras for years. It also served as a fun contrast to the distinctly garage-band quality of some ground-based telescopes, which regularly feature the scientific equivalent of roadies lugging amps or someone duct-taping a wonky microphone together and random people popping in to twiddle a connection or just test the air before the night starts.

On the night after our "significant meteorological event," the mission briefing was laced with a much stronger air of excitement. Part of it was just the accumulated tension of two dozen people crammed into a room and anxious to fly. I was also more than a little antsy. After waiting an entire extra day, on a weird sleep schedule, under a thick layer of cold clouds that occasionally spat snow, rain, and even hail into the California desert, wondering just how much turbulence SOFIA could tolerate riding through, and listening to Ravel's *Boléro* on loop after hearing Emily's symphony comment, my nerves were on high alert.

Things looked promising at first. If SOFIA runs like a symphony, the mission director of each flight is undoubtedly the conductor, learning everyone else's parts backward and forward and taking charge of bringing them all together for a successful performance. Our mission director reported that the weather was still lousy but apparently not full-on dangerous, and it looked like we might actually make it onto the plane and into the stratosphere. The whole group perked up at the thought of a successful flight and quickly cleared the room once the briefing concluded, yanking on coats and backpacks and juggling night lunch as they hurried outside. Swept along with the crowd, I was practically vibrating with excitement as I copied the people around me in grabbing earplugs, slinging a reflective safety strap over my shoulders—a safety precaution for crossing an active taxiway—and heading for the plane.

On board, everyone dispersed to settle in, stashing bags under seats and night lunches in a couple of dedicated refrigerators that, along with coffeepots and a few small microwaves, made up the plane's tiny galley. The temperature on the plane was already noticeably brisk; most of us left our coats on as we scattered to check out the telescope and instrument workstations, admire the back end of the telescope chamber where the infrared camera was mounted, and pace the length of the plane as we waited for the final word on whether we were flying.

That was the evening of the troubled cooling system, the threat of ice on the wings, and the frowning stuffed unicorn. When the final call— and second flight cancellation in a row—eventually came from the mission director, I understood along with everyone else that it was the right decision, but I was still disappointed. It had taken months of effort to get approved for this flight as a journalist and author, hoping to gain some firsthand experience with airborne astronomy as part of my research for this book. The SOFIA staff had already been immensely helpful, but I knew it would be difficult for them to simply stick me on another flight. I suspected the usual archnemeses of astronomy—weather and equipment problems—had just ended my only chance to fly on SOFIA.

o o o

By flying up to forty-five thousand feet, SOFIA does a pretty excellent job of getting above most of the earth's water vapor and giving us access to longer-wavelength infrared light; still, the more atmosphere we can put below us, the better.

Hubble and other space telescopes manage this by orbiting high above the planet, but putting a fully operational observatory into orbit quickly becomes an expensive and technically challenging proposition. A telescope is already a complex engineering achievement, and when that's combined with weight restrictions, launch capabilities, and designing something that can be operated remotely with little to no human intervention, a space telescope becomes an immense undertaking. Cranking out space telescopes of the same quantity and size as ground-based telescopes simply isn't practical.

Still, there are middle grounds we can reach that at least get close.

Airborne astronomy has traditionally referred to telescopes taken aloft by planes, but some astronomers have turned to balloons or even suborbital rockets to squeak above as much of the atmosphere as possible.

Astronomy balloons work at mind-blowing altitudes of around twenty-five miles, or about 130,000 feet. This far above the atmosphere, it's possible for us to observe gamma rays, X-rays, and ultraviolet light, and balloons have mapped the thin wisps of gas drifting between galaxies, picked up gamma rays from the remnants of supernovae, and served as the initial test beds for cutting-edge instruments that have gone on to fly on full-scale space missions.

If, like me, your first mental image when hearing *balloon* is of a kid's birthday party or a colorful hot air balloon—maybe with astronomers riding along in the basket—it's worth considering what the reality of balloon observing entails.

The actual balloons used for astronomy are massive—the largest are over 450 feet tall when inflated—and get sent into the sky carrying telescopes or detectors that are operated remotely by teams of astronomers who stay behind on the ground. While the detector itself can be controlled, the balloon is at the mercy of the winds at whatever altitude it's designed to float. As a result, launching one of these balloons involves a weird alchemy of wind currents, weather, timing, and teams scrambling to perfect and check and double-check their payload, the bundle of instruments and cameras (which can weigh several tons), before sending it into the sky.

Most U.S. research balloons are launched from a few primary sites in New Mexico, Texas, Australia, and Antarctica, all of which are equipped with balloon teams that specialize in this sort of work. On a good launch day, prevailing winds at several different altitudes—the ground, the intermediate layers, and the height the balloon will ultimately reach when it's "at float"—will all be cooperating, and the weather will be deemed suitable by a dedicated specialist working at the launch site.

The balloon itself—longer than a football field but as thin as a plastic bag—is rolled out along the ground on tarps. Attached on a line below the bottom of the balloon is a packed parachute, and down the flight line from the parachute is the payload, which waits for launch suspended above the ground from a tall mobile crane and held in place with a large release pin. Once the balloon is rolled out, the top fifth or so gets filled with helium, with an explosive collar fitted around the balloon to prevent the rest of it from filling. The balloon is held in place with a heavy spool to keep it on the ground, at least for now.

In principle, the actual launch of the balloon is a smooth and peaceful process. Once the helium fill is complete, the spool is released, and the balloon starts rising into the air. As it rises, it begins to lift the flight line behind it, and a member of the launch crew starts driving the movable crane to keep the payload carefully aligned with the balloon. Just as the line is pulled taut, the pin attaching the payload to the crane is released, and the whole thing soars peacefully away into the sky. The explosive collar around the balloon is eventually blown so the helium can expand to fill the whole balloon as it climbs. Once it reaches its target altitude, the balloon is at float, remaining suspended more than twenty miles above the ground for anywhere from ten to thirty hours as the astronomers on the ground happily observe using the telescopes and detectors sent up in their payload.

It's easy enough to describe, but it's also not hard to spot ways this process can go wrong. The launch alone can introduce some pretty substantial complications. A helium collar blown too early can cause a balloon to overfill, spring a leak, and sink back onto the launchpad, leaving it to sit there for days as helium slowly empties out. The launches are also strongly dependent on wind, and even very careful launch crews and weather experts have found themselves suddenly taking off in a random direction when the ground-level winds change. The timing of releasing the crane pin is also critical: too early risks dropping the payload, too late and the payload could slam into the crane arm or get yanked around on the line, and not releasing at all can be catastrophic. At the same time, there's only so much the launch crews can do. One balloon astronomer recalled seeing a launch crew member deal with a stuck pin by scrambling up the arm of the moving crane and kicking it loose. He was lucky—the pin released, and he was left perched on the crane as the payload rose away—but it was horrifying to think of what could have happened if the crew member had fallen, or worse, gotten tangled in the flight line as the balloon took off.

Eric Bellm recalled one infamous launch disaster during a balloon campaign he led at Alice Springs in Australia back in 2010. His team's payload was a telescope designed to observe gamma-ray light; during a previous flight, it had gathered excellent data on gamma rays from a nearby pulsar, and from Australia, they hoped to map the gamma rays coming from the center of the Milky Way, gather data on supermassive black holes, and keep watch just in case they spotted a gamma-ray burst from a distant dying star (the same high-energy flashes I'd been trying to explain in my thesis research). Working with an experienced balloon team and a payload that had flown before, Eric had decided to invite some media out to watch the unusual spectacle of a scientific balloon launch. Word spread, and along with several camera crews, a group of spectators arrived to watch the launch process, clustered behind a fence that was upwind of the launch site. Usually.

On the day of Eric's launch, the prevailing winds were in a different direction. As they prepared the balloon that day, drawing a line from the balloon through the payload pointed you directly to the fence. Still, this wasn't a source of concern to the local balloon crew; they'd launched in conditions like this before without a hitch. When the time came to launch, the balloon was released, the crane started to drive as it rose and started to tug on the payload...and nothing happened. The payload wasn't releasing, the balloon was still moving, and the crane kept barreling along underneath it, directly toward the fence and the area where the excited onlookers had parked their cars and gathered to watch.

Eventually, the crane ground to a halt at the fence, unable to drive farther, and the payload still hadn't been released. At this point, the balloon was still inexorably moving forward, with the payload swinging over the fence and sending onlookers scattering as they realized something was going very wrong. The strap connecting the payload to the crane snapped, sending the two-ton bundle of scientific instruments crashing to the ground and plowing through the fence on a balloon-dragged path of destruction. It knocked over one spectator's car, narrowly missed a second (that still had passengers inside), and cartwheeled into the air as people hurtled out of the way, pieces of equipment flying off as the whole payload disintegrated. By then, a launch crew member had started screaming "Abort!" into a walkie-talkie system, and the launch team cut the balloon loose, leaving the payload to smash into the ground and roll to a stop.

Eric and his team were left speechless, gawking at the shattered wreckage that had been their science payload and the shaken group of spectators who had been directly in its path. The whole debacle had taken less than a minute.

In an incredible stroke of luck, nobody was hurt. Unfortunately, the expectant media had their cameras running, so footage of the crash from two different angles quickly made it onto the nightly news and then YouTube. In the news report, onlookers were interviewed about their mad scramble away from the payload, while in the background, the dazed and dejected group of scientists picked through the remains of the payload. Sadly, it became clear that they wouldn't be able to salvage their equipment in time for another flight anytime soon, and the team flew home.

Getting the balloon up is tough enough, but it's also only half the story. At the end of a flight, it, or at least its payload, has to come *down* again and be successfully retrieved. In theory, another explosive element attached to the balloon will slice through the line and give teams control over when and where a payload drops (especially crucial when keeping an eye on prevailing winds and avoiding populated areas, protected airspace, or international borders). The parachute attached to the flight line is there for precisely this purpose; it deploys once the payload has descended into sufficiently dense air for the chute to operate and then detaches once the payload hits the ground. That said, this isn't always foolproof.

Once the payload is in the air, the worst outcomes of balloon flights are generally free-fall accidents, when the explosive sequence fails or the parachute doesn't deploy and the painstakingly built astronomical equipment winds up plummeting back to Earth and slamming its way several feet into the ground like Wile E. Coyote. On the other hand, a parachute can deploy but then never detach, which can end up dragging the payload for miles through the desert dirt or Antarctic ice. Usually, these incidents end similarly to Eric's launch crash in Australia, with teams of scientists searching through the wreckage with shovels in an attempt to at least salvage the truly irreplaceable bits of their equipment.

Successfully dropped payloads also have to be retrieved, which can be an adventure even in ideal launches. A payload launched in New Mexico eventually landed in what one team member succinctly described as "a canyon full of snakes"; the lead scientist was overheard wondering aloud whether "snake chaps" (thick leg covers that could in theory let the wearer wade into a sea of rattlesnakes unharmed) were an acceptable item to list in his National Science Foundation budget. The hassle of retrieval gets instantly worse when a payload winds up far from its intended landing point. While most are equipped with GPS trackers, it's still sometimes a surprise to see precisely *where* a balloon's payload has landed…or who might have gotten to it first. One team on their way to retrieve a balloon was driving down a remote one-lane road and passed a man driving a flatbed truck in the opposite direction; a closer look at the truck's contents revealed that he was making off with part of the payload. Another team supposedly launched a balloon from Australia that wound up crossing the Atlantic and landing in Brazil. When it came down, it landed across some power lines to a village and cut off the electricity. The team arrived months later to retrieve their equipment and stopped for a drink in the village, only to spot a chunk of the payload mounted over the bar.

A cousin of balloon astronomy is rocket astronomy, which works on the same basic principle as the balloons—toss an instrument up to a high altitude, take data, land, retrieve the instrument—but at a literally explosive pace. Stopping just short of full-on space-based efforts, teams will launch sounding rockets that fly in a giant suborbital parabola, carrying telescopes briefly into space for *just* long enough to capture a few minutes of precious data.

Arthur B. C. Walker II, a renowned aerospace engineer and early innovator in rocket-based astronomy, drew on experience with rocket launches from his time in the U.S. Air Force to design ultraviolet and X-ray instruments that could be used aboard sounding rockets. In the 1980s and 1990s, Arthur and his team launched payloads consisting of more than a dozen telescopes on several sounding rockets, recording the first high-resolution X-ray and ultraviolet images of our sun. This was in spite of the fact that an entire rocket flight would last, from beginning to end, about fourteen minutes, with only five of those spent in space and gathering data. Like scientific balloons, the rockets typically release their payloads to drop under a parachute and be retrieved by the team after the flight.

There are also, once again, only a handful of sites where these types of rockets can successfully be launched. One of these is the White Sands Missile Range in New Mexico (in the same gypsum sand dunes as White Sands National Monument, just west of Apache Point Observatory). Several colleagues described going through multiple rocket launches at White Sands and worrying about rockets getting aborted if they veered too close to population centers or the Mexican border. Sometimes the rocket payloads would also land in an area of White Sands that was somewhat less than accessible; parts of the missile range can potentially contain unexploded ordnance. Astronomers went through special safety training as part of their work on the range and were occasionally brought out to retrieve a payload in the back of a truck with a minesweeper walking ahead of them. Another team launched a rocket in Alaska in January and had the payload parachute fail, sending it a solid twenty-five miles off course from its expected retrieval point. Faced with minimal daylight and hostile terrain, the team wound up putting a financial bounty on the payload, awarded to whoever could track it down in the Alaskan wilderness. It worked: two summers later, someone found the moss-covered payload embedded in the ground (and claimed the reward).

Another launch site commonly used for rocket astronomy is the tiny island of Roi-Namur in the Kwajalein Atoll of the Marshall Islands, a remote cluster of Pacific islands near the equator with a military launch site that NASA can rent for scientific research launches. A challenge on Roi-Namur is ditching the rocket and landing the payload properly; with the island surrounded on all sides by environmentally delicate coral reefs, dropping anything from the launches into the nearby ocean is strictly prohibited. Rocket missions aren't generally quite as sensitive to wind as balloons and can typically launch at their scheduled time without any problems, but one colleague of mine, Kevin France, described waiting for five nights on Roi-Namur, raising and lowering the rocket and aborting takeoffs over and over due to winds that risked pushing the rocket onto the reef.

Another quirk of Roi-Namur is that the island has no cars; aside from a few service trucks and one police car, visitors rely on golf carts or their own feet to get around. To speed up their team's transportation around the island, NASA brought along a shipping container of disassembled cruiser bikes for the scientific teams to use. Kevin recalled being sent to build his bike upon arriving at the island. Team members quickly warmed to the bikes, decorating them with mission stickers and using them at all hours of the day and night to get to and from the launch site. The net effect was a bunch of NASA rocket scientists pedaling around a tiny remote Pacific island like overgrown twelve-year-olds out after curfew, bike lamps on and research gear strapped to their fenders or backs.

Airborne, balloon, and rocket astronomy are admittedly stretching the limits of what we might call ground-based astronomy, but I'm happy to lump them in with our earthbound telescopes. The data itself may be taken many miles above the planet, but the whole endeavor starts and ends on the ground, and the observers are either riding along or desperately tracking every precious second of the observations as they happen.

If we really want to push it, George Carruthers undoubtedly wins the "most impressive ground-based telescope" contest, albeit on a technicality. The Far Ultraviolet Camera/Spectrograph he designed was most assuredly placed on ground, but that ground was the surface of the moon, and his telescope operator was astronaut John Young, who operated the telescope during the Apollo 16 lunar landing in 1972. George was an early pioneer of ultraviolet detectors (he holds the patent for the first ultraviolet camera and also conducted early ultraviolet observations with sounding rockets) and built the diminutive three-inch telescope especially for use on the moon. It returned some exquisite data. George and his colleagues published several high-profile research papers on the results, including data on Earth's atmosphere and magnetic field thanks to, for the first time in history, being able to point a telescope back at our own planet. The little telescope also had its own quirks. The cable lines connecting the telescope to its battery kept getting tangled up in the astronauts' legs, and the battery itself had to be left in the sun to keep it warm. In the cold lunar temperatures, the lubricant that was meant to make the telescope easy to spin also started to freeze up, making it increasingly difficult for the astronauts to turn and point the telescope. Eventually, they were all but arm-wrestling the thing to get it to turn and then digging it out of the lunar dust to relevel it after each move, taking up valuable time and oxygen. Still, the telescope was deemed a success, and George later went on to launch a similar ultraviolet telescope for use aboard one of the Skylab missions in the 1970s.

The moon landings did offer us one additional bonus bit of astronomy. During Apollo 11, Neil Armstrong and Buzz Aldrin placed a special mirror on the lunar surface, designed to reflect lasers shot at the moon by telescopes. This means that by bouncing lasers off these mirrors, astronomers can directly measure the distance between the earth and moon to within a few millimeters; we learned from this that the moon is actually spiraling away from us at a rate of 3.8 centimeters per year. Apollo 14 and 15 added more mirrors during their missions, and Apache Point Observatory is still using them for lunar laser experiments today.

Stopping short of the moon, the most remote place where we can currently build and operate a telescope is the South Pole. The Amundsen-Scott South Pole Station, set near the geographic South Pole in the heart of Antarctica, is home to the South Pole Telescope, a ten-meter radio telescope that works at submillimeter wavelengths and serves as one element of the planet-wide radio interferometer used to image a black hole. If this sounds similar to the ALMA submillimeter observatory in the Atacama Desert, there's good reason. The South Pole itself is actually a high-altitude desert: the South Pole Telescope is over nine thousand feet above sea level, and there's very little precipitation at the site. Photos at the South Pole that look like snowstorms are usually capturing snow on the ground being blown around by high winds. The site can also *feel* even higher and more oxygen-deprived thanks to the earth's atmosphere being thinner at the poles, and the "feels like" altitude is actually reported on an information board at the station that lists daily weather conditions.

Astronomically, the South Pole is a spectacular site, but it also poses a particular challenge. Antarctic winters are, unsurprisingly, brutal, and personnel working there during the eight-month stretch of time when it's too dangerous for planes to fly to and from the continent are known as "winter-overs"; they're stuck on the continent until flights resume. It can be a long and trying stretch of time for anyone, and life is especially quirky at the relatively sparsely populated Amundsen-Scott Station (as opposed to McMurdo, the sprawling outpost closer to the coast). Roopesh Ojha, who has wintered over to work on the South Pole Telescope, described preparing for the trip and working through, in addition to his usual list of science tasks, details like exactly how much toothpaste or shaving cream he'd want to bring with him to last for the long months of the winter-over. That far inland, people also start missing all sorts of stimuli: the entire environment is sheer white (and, in the winter, dark), the food can take on a dull sameness after months of eating whatever's been shipped in and preserved, and even people's sense of smell will start to miss anything that isn't the icy, bone-dry Antarctic air. Roopesh remembered smelling bananas and fresh eggs for the first time after a winter-over when a new shipment of food arrived as well as stepping off the plane to New Zealand at the end of a trip and inhaling the rich scents of mud and plant life and fresh rain.

If you think that South Pole researchers can simply entertain themselves with cute baby seals and fleets of penguins, think again. Antarctic wildlife is primarily distributed in a ring along the outer edges of the continent, depending on the ocean water for food, which leaves the Amundsen-Scott Station almost entirely bereft of animal life. While the earliest explorers may have come to the continent with dogsled teams, the Antarctic Treaty System now prevents visitors from bringing any animals to the island. The effect can leave residents starved for a pet, *any* pet. Roopesh remembered a small green worm making it all the way to the station's kitchen, nestled in a head of lettuce. When the cook proclaimed in surprise that the little worm was alive, half the people eating lunch hurried over to get a look; they eventually dubbed the worm Oscar and kept it alive in the galley for a little while. The station also has a Roomba, affectionately named Bert, tasked with keeping the halls clean even during the long winter months when the station is down to a skeleton crew. In a place so remote that it barely even feels like Earth, the scientists working there are happy to latch onto whatever familiar human comforts they can, even if that comfort winds up being an intrepid worm or a robot pet.

o o o

I had one last nugget of hope left for SOFIA. Months before my ill-fated February flights, I'd submitted an observing proposal to the telescope in the hopes of answering one of my long-standing questions about red supergiants. I was still curious about the dusty shrouds of material surrounding these dying stars—how they formed, what they were made of, and whether they held any clues about the stars' eventual fate—and I knew SOFIA was uniquely capable of observing the faint, long-wavelength infrared light that this dust emitted. If I got SOFIA time, I'd be able to get a unique look at the evolution of these stars, one that no other existing telescope could manage…and I might get the chance to fly again.

When I learned my proposal had been approved, I jumped at the chance to get on another flight in July. This time, SOFIA would be flying out of the U.S. Antarctic Program's base in Christchurch, New Zealand, and I'd be on the plane as an astronomer in my own right, with one of the stars that I was studying slotted into the flight plan.

Frantically rearranging my summer travel plans, I booked a masochistic series of flights that meant I could rush to New Zealand straight from a conference presentation I'd just given in France during a brutal heat wave. I already traveled a lot for work, but flying halfway around the globe *to then fly on another plane* was a little extreme even for me. For years, Dave had joked that with the way I traveled, I could easily be concealing a secret career as a CIA agent, using "Oh, I have an observing run" as a convenient cover story to disappear to South America or Australia or some other far-flung locale for top secret hijinks. Scrambling to book a last-minute trip to a different hemisphere probably did nothing to disabuse him of this notion (though hopefully putting it in writing now definitively confirms that I'm not a CIA agent) (*or am I?*).

I arrived in Christchurch after three solid days of traveling, on a personal time zone that could only be described as "wrong," swapping France in summer for southern New Zealand in winter and sporting what was probably an alarming amount of enthusiasm as I marched gleefully through the frigid air. The whole experience was recognizable as observing on steroids: the long trip, the remote locale, the wonky time shifting, and propelling the whole adventure with some slightly manic excitement that would most likely wear off at precisely 3:00 a.m. during the observing run.

This time, the weather and the plane were both cooperating, and after egress training and the mission briefing, we all once again grabbed our backpacks and night lunches and reflective safety vests and belts and made the trek across the tarmac to where SOFIA was waiting, fueled and ready for takeoff and backlit by a blazing display of sunset colors. We'd be flying a funny inverted triangle that took us south over the Southern Ocean (and within spitting distance of the Antarctic Circle), looped east, and then flew north again, observing the center of our own Milky Way, the remnant of a star that had exploded several decades earlier, and a few not-quite-dead-yet stars, including mine.

As the only first-time passenger on the flight, I jumped at the offer to sit in the cockpit for takeoff. Riding behind the pilots and next to the flight engineer, I sat on my hands (not wanting to so much as nudge any of the numerous switches and buttons around me, even by accident) and grinned as we taxied through the bustling Christchurch airport, the pilots chatting with air traffic control to ensure we were in position to take off right on time. I'd been on enough plane flights at this point to savor the unique view of the runway lights stretching directly ahead, and I vibrated with excitement as the plane tipped up and the lights dropped away. I was finally doing this! The cockpit lights were dimmed for takeoff, and as we ascended from the airport and into the skies over New Zealand, a wealth of southern stars started fading into view.

We quickly climbed to forty-three thousand feet, a mile higher than most commercial planes fly. At this altitude, a colleague had pointed out, we'd be well into the stratosphere and could justifiably refer to ourselves as "stratonauts." I gawked out the window, grinning like a little kid and working hard to imprint the whole experience on my memory. I stayed in the cockpit until the flight engineer opened the telescope chamber—as promised, there wasn't so much as a wiggle of turbulence to indicate that we'd just opened an enormous door in the side of our plane—and then headed downstairs to join the rest of the crew.

The plane was loud, just as I'd been warned; I popped in earplugs and sealed myself into a weird sensory-deprivation bubble while I wandered around, broken only when I grabbed a headset to chat with people. Decibel level aside, the atmosphere on the plane was surprisingly quiet and calm. The telescope and instrument operators were plugging away, and other team members were microwaving dinner or busting out snacks. It was also, as promised, chilly; I tugged on a winter hat, zipped my coat up to my chin, and wished I'd thought to pack gloves. I made some tea in my trusty lockable mug and wandered to the rear of the plane to watch the back end of the telescope itself, protruding from its sealed chamber. As the rest of the plane bumped through almost-imperceptible turbulence, the telescope appeared to bob and weave ever so slightly, floating on its massive ball bearing to stay perfectly steady as it observed.

Like any other observing run, when everything was underway and going smoothly, things were…well, not anticlimactic exactly—I'd been looking forward to this flight for too long to not be perma-excited—but bizarrely *normal*. I made myself comfortable at a work station, took some notes on the flight, reheated the Thai takeout I'd brought for dinner, refreshed my memory on the details of my observations coming up in a few hours, and studied the data I hoped to see. I was still running on adrenaline but had also gotten about six hours of sleep in the past three days, so I suspected I'd need to build in a nap at some point during the flight.

Even though fatigue eventually set in, by the time my leg of the flight came up—a whopping sixteen minutes when SOFIA would be pointed at one of my red supergiants—I was awake, on a headset, clutching a fresh mug of tea, and hovering behind the instrument station (where the status unicorn was fortunately smiling away). On a plane or on the ground, there's always a little flash of nervousness when using a new telescope for the first time or working in a new subdiscipline or wavelength regime, and this time was no different. What if I'd screwed all this up? What if my observing idea was terrible and I was about to waste everyone's time getting data that would prove useless? It may have been a silly concern; a committee of astronomers had reviewed and approved my proposal, and I'd prepared furiously for the observations. Still, it certainly didn't feel silly when standing on a plane full of people who were all here—literally forty-three thousand feet high and in a very specific stretch of sky off the south coast of New Zealand, in a plane with a hole in it—because I'd said I wanted this telescope on this airplane to observe this thing. I watched the instrument tip a bit as it moved to point to my star and had a stomach-dropping moment of "please don't have screwed this up."

Happily, everything went flawlessly. The instrument scientists locked onto my star immediately, and the data began rolling in so quickly, I was almost startled. I hadn't screwed up; the coordinates had pointed us perfectly to my star, and it wasn't invisibly dim or horrifyingly bright. I had never worked with this kind of infrared light before, but the examples I had studied roughly matched what was showing up in the raw, unprocessed data: a thin white line arced across the center of the instrument panel's screen with little faded patches signaling that there were fingerprints of stellar chemistry embedded somewhere in that collection of light.

A few hours into the flight, I got permission to head back upstairs; I wanted to ask the pilots and flight engineer some questions. We were still zooming south, and the interior lights were on in the cockpit as the four of us kept an eye on the plane and chatted about the ins and outs of SOFIA... until one of the pilots cut in.

"Do you mind if we kill the lights? We think there's some aurora over there."

I nodded, the cockpit went dark, and I stopped breathing.

I'd never seen the aurora. I'd *dreamed* of the sight since I was a small child but had never once spotted them. It had turned into something of an obsession; the northern lights had made it down to Boston while I'd lived there, and I'd taken several commercial flights that skirted close to the North Pole, but I'd never once actually caught the aurora with my own eyes. Now, suddenly, here they were: massive pale green curtains, curling and waving and coalescing in a strange motion unlike anything I'd ever seen. We were practically *in* them; they filled such an immense swath of sky that it was hard to know where to look from the panoramic cockpit windows, flickering over my head and rippling and winding back on themselves straight ahead and forming an almost eerie glow low on the horizon.

The pilots called down to the rest of the plane and switched off some of the forward lights, where seats had been set up for people to rest and sit during takeoff and landing. Incredible as the aurora were, they weren't an entirely unexpected sight (they showed up with some regularity on SOFIA flights this far south), and everyone still had a job to do as the telescope kept observing. Still, as the flight went on and the aurora continued glowing brighter and brighter outside the window, people began drifting forward in ones and twos, crouching down to peer out the plane windows and briefly popping out earplugs to shout and geek out to one another about the astonishing light show happening outside. The SOFIA veterans had all seen the southern lights before, but they still warranted a few minutes of admiration, and the other observers on the flight were delighted. The green glow, we knew, came from oxygen atoms in our atmosphere being bombarded by charged particles from the sun, and the science behind the spectacle wasn't lost on any of us as astronomers. The aurora are, after all, caused by a star.

I was in the cockpit of a flying NASA observatory, speeding south through the Antarctic stratosphere. The aurora were arcing and dancing around us. In the back of the plane, a telescope was preparing to point at a star of my choosing, gathering data that would help me unravel one of the many mysteries of the universe that I'd dedicated my life to exploring.

It briefly occurred to me that the aurora might, possibly, have a slight impact on the quality of the data. At that moment in time—astronomer or not—I couldn't bring myself to mind.

CHAPTER NINE

THREE SECONDS IN ARGENTINA

Astronomers can sometimes be saddled with the typical stereotype of robotic scientists who lose the ability to see romance and beauty in favor of dryly spouting factoids or studying zeroes and ones. I discovered how thoroughly my fellow scientists defied this stereotype when I started talking to them about my eclipse plans.

Like so many other people, I would be seeing my first ever total solar eclipse in August 2017, when the moon would perfectly block the sun for a huge swath of the United States. My colleagues and I were all fully conversant in the nuts and bolts of the eclipse: the date and time and place, the safety tips warning us against looking directly at the sun without eye protection even when it was 99 percent covered, the white blaze of the corona (the outermost stretch of the sun's atmosphere) that we knew would appear at the moment of totality, and so on. Still, when I raised the subject, most people's first instinct was to talk about the beauty, the emotion, the almost spiritual ethereality of the whole event.

Astronomers' understanding of the mathematical poetry and scientific elegance behind an eclipse seemed to only deepen their ability to soak up its beauty. Observers who had seen eclipses before—either for research or for sheer love of the heavens—waxed on about the haunting quietude that settled over them at the moment of totality, the meditative calm of feeling at one with the solar system, the arresting sight of the white blaze of the solar corona, and how they really took the two and a half minutes of totality to drink in the exquisite beauty of the universe.

My own reaction proved to be a bit less Zen.

For the 2017 eclipse, I joined some colleagues in Wyoming who were organizing a kind of science tourism event, inviting a group of two hundred eclipse enthusiasts to view the event from the Jackson Hole Golf & Tennis Club in Jackson, Wyoming. I tagged along as one of the special guest speakers who would give science presentations to the attendees during the evenings. On the morning of the eclipse, we scattered along an open stretch of the golf course, the sky gloriously blue and clear over our heads and the spiky peaks of the Grand Tetons jutting up behind us to the west as we all pointed ourselves toward the sun like geeked-out flowers. A veritable forest of viewing devices sprouted up from the group: everyone was clutching a pair of dark-tinted eclipse glasses, but attendees were also busting out shielded binoculars, every possible stripe of digital camera, small telescopes with solar filters, and setups that could project images of the sun through pinholes and onto screens.

Many attendees were eclipse veterans—I talked to someone who told me this was their twentieth eclipse—but my family and I were first-timers. My friend Doug Duncan, who was organizing the event, had agreed I could have my family join me. He'd perhaps underestimated that for my family, this meant inviting a horde of sixteen people (Dave, my parents, my brother and his family, and a variety of aunts and uncles and cousins) who were all as uncontrollably excited as I was about seeing their first total solar eclipse. Most of the family had traveled via plane or car or camper van from Massachusetts to join me at this particular spot, banking an entire long and expensive trip on the hope that we would have good weather at precisely 11:35 a.m. in our chosen spot.

The sense of anticipation grew as the sun slowly disappeared, imperceptible for a long time save for the dropping temperature, a slight funny shift in the light, and the fact that every image being projected of the sun was slowly morphing from circle to Pac-Man to croissant. The group of enthusiasts on the field was cheerful and eager to chat, encouraging anyone passing by to look through this special telescope setup or that filtered set of binoculars. All around Wyoming—all around the *country*—people had flocked to the path of totality. People were jamming highways from Oregon to South Carolina, getting their hands on eclipse glasses or sufficiently dark welding goggles, or fashioning pinhole cameras out of whatever they could get their hands on (colanders, Ritz crackers, cheese graters) so they could watch the sun slowly vanish. The eclipse blanketed social media and popped up in almost every news and radio program that aired that week. It was delightful, I thought, to see so many millions of people excited about astronomy. It felt like a massive, countrywide observing run.

In the final moments before totality, the excitement became unbearable. Darkness was starting to descend, and everyone had their eclipse glasses on and their faces pointed skyward. Several people gasped and turned as, behind us, the Grand Tetons suddenly plunged into darkness, the first hint of totality's shadow speeding toward us at over two thousand miles an hour.

The light on the grass around us began to shimmer noticeably. "Shadow bands!" Doug called over the field, recognizing a particularly spectacular display of astronomical seeing, the same thing that makes stars twinkle. Just before and after a total eclipse, the turbulence of Earth's atmosphere refracts light from the narrow sliver of visible sun, making the final weak hints of sunlight undulate like ripples at the bottom of a pool. The whole group began to murmur and chatter and then shout, building to a crescendo of cheering and applause as totality hit and two hundred pairs of eclipse glasses were triumphantly torn off.

I was immobilized for a brief moment, gawking at the sky. An astoundingly perfect black void sat where the sun had been, surrounded by a jagged white nimbus of light that nearly brought me to tears. This was the solar corona, the hot outer edges of the sun's atmosphere that drive a flood of particles into space and generate a phenomenon known as a stellar wind, a key property of how our sun and other stars evolve. I had studied this particular aspect of stars for almost my entire life, using a dozen of the best telescopes in the world, but this was the first time I could see a star's wind with my own naked eye. Around us, the sky was a strangely uniform dome of sunsets in every direction, and the warmth of sunlight had been replaced by an almost primal up-the-neck chill. It felt like the planet itself had been put on pause at this particular place and moment in time, a frozen moment of "*look*."

According to my colleagues, this was supposed to be a moving and spiritual experience for the full duration of totality. Clearly, they were inherently calmer souls than I was, or at least more patient and restrained, because upon sight of a total eclipse, I proceeded to lose my ever-loving mind for the next two and a half minutes.

I couldn't figure out where to look. I obviously had to keep gawking at the eclipse, but I also clearly had to grab binoculars to see it up close and look through the spotting scope my dad had set up and wheel around in a crazy circle to take in the 360-degree sunset. I began pinging around my assembled family members like a freshly launched pinball. Were they looking? Through binoculars? Through the scope? At the sunset? *Was everyone looking??* (Yes. Yes, they were.) In the binoculars, I spotted Mercury just to the lower left of the sun—a planet we hadn't been entirely sure would be visible when discussing the details of the eclipse the night before—and my shout of "YOU CAN SEE MERCURY!" carried across the entire field and possibly the entire state of Wyoming. I ecstatically hugged Dave (and then hurled binoculars into his hands because *look, there's a glowing loop of plasma at the edge of the sun!*). I hugged my parents. I probably hugged strangers. When the sun reemerged, a momentary diamond ring as the first brilliant hint of the solar surface started peeking out from between the moon's mountains, I whooped and cheered again with the rest of the group, shouted "ECLIPSE GLASSES BACK ON!" (still at top volume) and then shot across the field to share my elation with Doug. As it turned out, a documentary crew was filming our group during the eclipse, and they caught me bouncing excitedly in place in the immediate aftermath of totality, out of breath like I'd just run a race and gasping that it had been "the fastest two and a half minutes of my *life*."

o o o

Total solar eclipses can be some of the toughest but most rewarding events to observe in astronomy. They give us a unique opportunity to study everything from the sun itself to complex theories of gravity and spacetime. Similar events also happen on a much smaller scale, known as transits or occultations rather than eclipses but following the same general principle. Observing the planet Venus as it passes in front of the sun can help us learn more about Venus's atmosphere and how to identify planets around other stars. Planets and asteroids in our solar system will even sometimes pass briefly in front of distant stars, giving us a few fleeting moments to study how the star's light changes and suss out new information about planet atmospheres, asteroid shapes, and how our solar system works.

The trick with all these—just like with the 2017 solar eclipse—is actually getting into position to catch them. Observing eclipses can be a herculean undertaking. Any eclipse-like event, whether it's the moon passing in front of the sun or a tiny asteroid momentarily blocking light from a star, is effectively casting a shadow on the earth, and that shadow only covers a small part of our planet. This means that unless we get terrifically lucky and have a top-notch telescope in precisely the right place, observers have to travel to the shadow. This is why so many people flocked to a narrow strip of the United States stretching from Oregon down to South Carolina in 2017 to catch the last total solar eclipse: the eclipse path marked the shadow cast by the moon on the earth as it passed between us and the sun.

For astronomers who make a career out of studying these sorts of events, this can mean taking on immense and equipment-laden expeditions to chase cosmic shadows all over the place. Since these events are most likely not passing over an existing observatory, we have to bring the observatory to them. This turns the act of observing into a sort of traveling circus, using equipment that can be packed up and hauled to wherever the shadow might land.

Where the shadow might land can also be hard to pin down in its own right. Astronomers have a pretty good mathematical grip on the relative motion of the earth, moon, and sun and can predict the timing and shadow position of solar eclipses perfectly. However, the math gets messier and more uncertain when we're dealing with, say, a distant asteroid with only rough constraints on its size, mass, distance, and orbit passing in front of a faint background star. A research group hoping to see that particular asteroid pass in front of that particular star may need to dispatch multiple teams to different potential sites and observe over a broad range of dates. Eclipses are also under no obligation to pass over ideal observing sites with cooperative weather and sky conditions. An astronomer could expend immense effort to reach precisely the right place at exactly the right time, only to see their plans foiled by a single ill-timed cloud.

o   o   o

The phrase *scientific expedition* tends to conjure up mental images of turn-of-the-century explorers doggedly trekking through the unexplored wilderness with only a team of horses, a shotgun, and their wits. It's true that eclipse observations have included some famous historical adventures, but it's also equally true that these types of adventures persist even today.

Eclipse expeditions from previous centuries have filled entire books, including David Baron's *American Eclipse,* recounting a worldwide Gilded Age scramble to study the solar eclipse that crossed North America in 1878. The most scientifically famous eclipse observation is undoubtedly Arthur Eddington's expedition in 1919 to test Albert Einstein's theory of general relativity. According to Einstein, the sun should "lens" background stars as it passes in front of them, bending their light thanks to the effects of the sun's gravity on space-time and causing them to appear at a slightly different place in the sky. The problem with testing this theory is that normally, the sun is so bright, it outshines the light of all the other stars in the sky. An eclipse would solve this, handily blocking the sun and allowing Arthur Eddington to measure the position of nearby stars. He just needed to pack up the cutting-edge astronomical equipment of 1919 and get into the path of the eclipse.

In 2017, many normally quiet corners of the United States found themselves overrun with astronomy enthusiasts as they crammed themselves, their cars, and their cameras into the path of the solar eclipse. Still, a traffic jam in Wyoming isn't quite as rough as what Eddington's 1919 trip entailed. More than two months before the May 29 eclipse, he sailed from England to Príncipe Island off the coast of western Africa, toting a huge telescope borrowed from an observatory in Oxford and a slew of fragile glass photographic plates. The equipment was set up and ready weeks ahead of time, and the team held its breath on the day of the eclipse as torrential rain and thick clouds blocked out the sky all morning. Luckily, just before the eclipse, the skies cleared, and they were able to take a few precious photographs that confirmed Einstein's theories.

Not every historical observer got this lucky: wonky equipment or badly timed clouds can be all it takes to stymie our observations, and when someone has traveled for months in the hopes of a few minutes of data, this can be frustrating to say the least.

Eighteenth-century French astronomer Guillaume Le Gentil is infamous for what may be the most epic failure in the history of eclipse observing. He set out for Pondicherry, India, to observe the 1761 transit of Venus (where the planet passes in front of the sun) and ran into problems almost immediately. War between France and Britain broke out midtrip, and with Pondicherry captured by the British, the French crew of Le Gentil's ship decided to bail on India and dock in Mauritius, off the east coast of Madagascar. Le Gentil gamely attempted to observe the transit from the bucking deck of the ship, but his results were useless.

Venus transits are extremely rare, but they also happen in pairs. Each pair of transits occurs less than once every hundred years, but the two transits in each pair are separated by only eight years, and the 1761 event was the first in its pair. Knowing he'd get one more shot in 1769, Le Gentil decided to simply hang out in the Indian Ocean for eight years and take another crack at Venus during the next transit. This time around, things started better. By 1768, Pondicherry was back under French control and welcomed Le Gentil enthusiastically, and he spent more than a year setting up an observatory. The next transit was due on June 4, 1769, which dawned stormy, with thick clouds blanketing the sky for the entire duration of Venus's trip across the face of the sun (and clearing immediately afterward).

Depressed and ill, Le Gentil dragged himself home to France to discover that he should have probably sent a letter sometime in the intervening eleven years. Knowing only that he'd shipped off to India and never returned, his friends and family back in France had decisively moved on: he'd been declared legally dead, his wife had remarried, his heirs were bickering over his estate, and he'd been booted out of the French Academy of Sciences (which had sent him on the trip in the first place). The whole ordeal officially cemented Le Gentil's status as the record holder for "worst business trip ever."

o o o

Today, airplanes and cell phones make travel considerably easier, but solar astronomers still have to crisscross the planet every few years to put themselves into the paths of eclipses. Shadia Habbal, who uses eclipses as an opportunity to study the sun's wind and magnetic fields, has led ten solar eclipse expeditions to locations all over the globe. By following the eclipses, she and her team get repeated opportunities to study the outer regions of the sun precisely when they're easiest to observe, thrown into sharp relief thanks to the sun itself being completely blocked by our moon.

Shadia has certainly hit her share of bad luck with weather and observations: a snowstorm in Mongolia in 1997, a terribly timed cloud right at the moment of eclipse in South Africa in 2002, and a sandstorm in Kenya in 2013. She's also had to wrangle considerable logistical challenges. Sites for professional eclipse observations are chosen through a combination of the eclipse path and the anticipated climate in the area, both meteorological *and* political, as Shadia discovered for a solar eclipse in 2006. The path of that eclipse arced through northern Africa and was ideally observable from southern Libya, which could have potentially posed a challenge. Luckily for the astronomers, in early 2004, the U.S. State Department ended what at the time was a more than twenty-year ban on citizens traveling to Libya, making it a workable site for a large eclipse expedition. Eclipse observing teams actually wound up receiving aid from the Libyan army to transport equipment to the southern part of the country, even going as far as setting up a dish for internet at the research campsite in the middle of the desert.

Another solar eclipse in 2015 passed over the Arctic Circle, and Shadia's team chose Svalbard in northern Norway as the ideal site for the observations. With their base of operations in a beautiful but high-walled valley, they carefully scouted out a spot where the low Arctic sun would manage to peek over the mountains and be visible during totality. On that trip, in addition to the visibility and weather and equipment, Shadia's team had to worry about polar bears. While black bears at other observatories can usually be dealt with by simply giving them a wide berth, polar bears are a different matter altogether. Several members of the 2015 eclipse team were trained for target shooting and issued a rifle upon arrival in northern

Norway, and while working out in the exposed heart of the vast snowy valley, someone was constantly keeping watch. Still, even fear of polar bears was able to take a back seat to what became an extraordinarily beautiful eclipse, with the solar corona illuminating the snow-blanketed valley walls all around them. The town where Shadia's team was staying closed everything down fifteen minutes before totality so that every resident could go out and watch the eclipse.

Shadia also recalled a similar local reaction from another expedition that otherwise couldn't have been more different in terms of location. Rather than the Arctic Circle in March, she was in French Polynesia in July for a 2010 eclipse that passed directly over the tiny atoll of Tatakoto. Shadia, who speaks fluent French, gave a lecture to children at the primary school on the island, explaining the importance of wearing eclipse glasses along with the crucial step of removing them as soon as totality hit so they could fully take in the blacked-out sun and blazing white corona. The school's headmaster later approached Shadia and proposed a plan: if she could tell him precisely when totality would hit, he'd ring the church bells at that moment and spread the word around the island in the meantime that people should remove their glasses when they heard the bells so they could admire totality. The plan worked flawlessly, and the roughly two hundred and fifty residents of Tatakoto were able to enjoy a spectacular eclipse.

The raw human appeal of these events was familiar to me after the 2017 eclipse. It reminded me of how the entire town of Jackson in Wyoming had turned itself out for the event, with banners hanging on streetlamps, eclipse-themed beers and specials served in restaurants, and eclipse-specific jewelry on display in stores. The moment of totality is pretty much impossible to miss if you're in the path of the eclipse, and the universal bond shared by everyone standing underneath the sun as it momentarily disappears is palpable.

o o o

Solar eclipses are dramatic and spectacular and well-studied events: thanks to orbital math and carefully measured distances, we can predict solar eclipse times and locations down to the second. However, as astronomers move to smaller events—say, a small asteroid passing in front of a background star—everything gets a little fiddlier and a little harder to predict. The timescale of these sorts of brief star-blocking events, known by their observers as occultations, gets shorter: small and fast-moving asteroids may only block the light from a star for a matter of seconds. Exactly when and where the shadow gets cast also moves from the realm of certainty into a maze of probabilities, with different time and location possibilities laid out depending on how well we think we know the distance, orbit, shape, and size of the asteroid.

The stakes also go up accordingly. Solar eclipses happen somewhere on Earth about once every eighteen months on average, but the odds of any one particular asteroid crossing neatly in front of a sufficiently bright background star, in a place that's feasible to reach with teams and equipment, quickly diminish to the point where every occultation functions as a one-time opportunity.

In 2014, astronomers discovered a little rocky object named 2014 MU69 in the Kuiper Belt, the wide ring of small rocky bodies orbiting our sun in the outer reaches of the solar system. The object itself wasn't especially remarkable save for one key fact: we'd be able to pay it a visit in 2018. The New Horizons space probe had launched in 2006 and taken a nine-year trip to fly past Pluto and take the first close photographs of the dwarf planet's surface. The flyby was a resounding success, and after zooming past Pluto, New Horizons was still flying. Anticipating this, astronomers had searched for a suitable object in the Kuiper Belt for New Horizons to aim for and came up with 2014 MU69. With a few course corrections, New Horizons would be able to fly past 2014 MU69 beginning in late 2018 and get a close-up view on January 1, 2019. However, in advance of this visit, astronomers wanted to gather as much information as possible about this odd object. We would only get one shot at the flyby with New Horizons, so gathering as much early data as possible would let us take maximum advantage of the visit.

This challenge was where occultation observers got the chance to shine. Astronomers knew that 2014 MU69 was small (observations would eventually measure it at twenty-two miles long) and very dim (as a rocky body, it emitted no light of its own and was only discovered thanks to its weak reflection of the sun's light). However, orbital math suggested that 2014 MU69 would be passing in front of three different stars on July 3, 10, and 17, 2017, and teams were dispatched across the planet to try and catch the occultations, scattering several dozen mobile telescopes throughout South Africa and Argentina. That July, multiple teams hit pay dirt, catching fleeting shadows cast by 2014 MU69 at their chosen observing sites that made it possible to further nail down the little object's orbit as well as its shape.

Even SOFIA joined in the fun. Airborne observatories offer a distinct advantage in this business by being portable and able to fly into the path of the event. Jim Elliot had demonstrated this previously with his research on Pluto's atmosphere and the rings of Uranus using the Kuiper Airborne Observatory; both were discovered thanks to the planets, atmospheres, and rings occulting stars behind them. SOFIA devoted the July 10 flight of its 2017 Christchurch deployment to chasing the 2014 MU69 shadow. With a flight path and schedule that were kept accurate down to the second, the SOFIA team managed to catch a brief glimpse of the occultation and helped researchers learn more about the edges and surrounding environment of the mysterious object.

Collectively, all the key observations of 2014 MU69 occulting nearby stars couldn't have added up to much more than a minute or two of data, but it was enough. The dozens of telescopes and team members and travel hours combined to reveal that 2014 MU69 had a funny oblong shape; it looked like it could be two objects stuck together or orbiting each other incredibly closely. The data helped the New Horizons spacecraft to fine-tune its flyby plans, and the first pictures from its approach immediately revealed that the teams' work had been correct. The spacecraft sent back an exquisite high-resolution image of 2014 MU69, looking like a pair of snowballs stuck together to make a snowman. At over four billion miles away from Earth, it's the most distant object we've ever visited with a spacecraft.

With all the scrambling for science, it's easy to see the many ways that logistical challenges could throw roadblocks into the paths of astronomers trying desperately to chase down a tiny starlight shadow. The timing alone is brutal; 2014 MU69 may have been particularly short, with occultations lasting at most a few seconds, but it's far from alone when it comes to lengths of most occultations. Teams will travel thousands of miles and set up observations deep in the middle of nowhere, planning far ahead and giving themselves plenty of buffer, but a single math error, a momentary equipment malfunction, or an unfortunate cloud can render the whole adventure data-less.

Locations can also vary wildly when it comes to accessibility. Occultation observers' travels have ranged from "drive up to Palomar Observatory because an occultation is passing directly overhead" to long treks out to small Swiss mountain towns or remote islands scattered across any ocean you can name. The remote locales can make it difficult to deal with even fairly routine equipment problems. At the same time, events like this can be met with the same level of cheer and enthusiasm that greets solar eclipses. Larry Wasserman recalled the commotion that ensued in Comodoro Rivadavia, a city of about 180,000 in southern Argentina, when the 2014 MU69 occultation teams arrived in 2017. The observations had been covered in the local newspaper, and in a city that didn't get many English speakers, Larry would get stopped by locals while he was running errands or grabbing lunch and asked if he was part of the astronomy expedition that had come to town. On the night of the observation, streetlights in the city were shut off, and a major national highway was blockaded and closed down for hours to try and minimize light pollution. The team had been worried about gusty winds shaking their portable telescopes and making it hard to get high-quality data; to prevent this, truck drivers positioned their vehicles alongside the observing setup to serve as wind breaks. Life in Comodoro Rivadavia temporarily screeched to a halt that evening, all to allow a few moments of observations as 2014 MU69 blotted out a nearby star.

Hearing these stories had me reflecting on a funny irony of astronomy. One fundamental fact that every single professional astronomer can unanimously agree on is that astrology—the dubious superstition of divining information about human behavior and earthbound events based on the apparent positions of stars and planets—is definitively, unquestionably, not real. Even a cursory glance at the science can quickly establish that the apparent motion of Mercury ("Mercury in retrograde" is a simple trick of perspective thanks to the relative orbital speeds of Mercury and Earth, which can create the illusion that Mercury is sometimes moving backward in the sky) or which constellation is sitting invisibly behind the sun when we're born (because yes, the constellation corresponding to your sign isn't even observable when you're born; it's up during the daytime and impossible to see!) have absolutely no bearing whatsoever on humans' inherent natures, habits, and destinies. I have never met an astronomer—the people who are, by education and training, definitive experts on the night sky—who puts any stock in astrology.

At the same time, an amused colleague once pointed out to me that as astronomers, we, more than anyone else, actually *do* have our lives dictated by the stars, albeit not in a mystical or metaphysical way. A broken telescope or windy night may postpone a PhD defense, impacting a young researcher's career and life plans. Bad timing during an occultation trip can mean the difference between a celebratory newspaper headline or a glum trek home. Le Gentil certainly had the course of his life changed thanks to a Venus transit. Less tangibly, the sight that had driven two-year-old me to gawk at Halley's Comet in my backyard had clearly propelled me to MIT, where I'd met

Dave; to grad school in Hawaii, six thousand miles from where I'd grown up; to a career that had taken me all over the planet; and to a golf course in Wyoming where more than a dozen family members had come together to see their first solar eclipse. It may be chance and choice—the statistical wiggles of the universe combined with the wacky decisions of those of us who have decided to lash our careers and fates to the cosmos—but while astrology itself may be bunk, astronomers are well acquainted with just how powerfully the skies can affect our lives.

o o o

As activity wound down on our Wyoming golf course following the 2017 eclipse, my family and I milled around and savored the final moments. We checked out the sun occasionally through our glasses as it slowly finished reemerging from behind the moon and chattered about what we had seen. Several people were already asking about April 2024, when the next total solar eclipse to hit the United States would be tracing out a path that arced over the eastern swath of North America from Mexico and Texas to Maine and Nova Scotia. It would be another short drive or flight away for the veteran or newly minted eclipse enthusiasts in the crowd.

That day was, when stopping to think about it, a truly epic confluence of events. Hundreds of humans had all converged on this particular spot in Wyoming, a result of years of planning and packing and preparation. There hadn't been any clouds blocking our view. Perhaps most shocking was the eclipse itself. It's easy to miss this fact, but Earth's total solar eclipses are actually a stunning quirk of fate, a party trick of our solar system that likely sets us apart as a rarity even among the many thousands of planets that have been discovered around other stars. We're still actively learning about planets around other stars and the moons that might circle them, but one thing is certain: the odds of achieving such a thoroughly exact match between the projected size of our solar system's central star, 93 million miles away, and our perfectly round little moon, only 240,000 miles away, are very small indeed. In some distant future, if we wind up in some sort of interstellar travel network with communities of intelligent friendly aliens, Earth's solar eclipses would probably function as a planetary tourism draw in the same way that the Grand Canyon draws people to Arizona.

There was also another layer of incredible serendipity thrumming under the surface on that August day during the eclipse, although only a relatively small handful of people knew about it at the time. It was the reason I'd pulled out my phone and started poking at the internet shortly after the eclipse had completed, admiring heaps of totality photos on social media from astronomer friends scattered around the country but also keeping one eyeball on my email. It was why a gaggle of scientists attending the High Energy Astrophysics conference that had been strategically placed in Sun Valley, Idaho, were simultaneously enjoying the sight of a total eclipse on a work trip and bemoaning the struggles of rural and overtaxed internet. Just four days before the eclipse, a type of signal unlike anything we'd ever detected before had arrived at Earth, streaking through the cosmos from 130 million light-years away and throwing a handful of astronomers into a frantic and quiet chaos.

CHAPTER TEN

TEST MASS

Four days before the 2017 solar eclipse, on the afternoon of August 17, Dave and I were waiting in line at a neighborhood ice cream shop. I was busy pondering the available flavors, texting back and forth with my dad to wish him a happy birthday, and gushing to Dave over how excited I was to see my first total eclipse with him and the rest of the family in just a few days.

In the midst of texting and surveying my ice cream options, an email arrived from Phil Massey. We were still working together on red supergiants, and I knew a mutual collaborator of ours was observing that night in Chile for one of our projects. Phil was passing along an urgent email from another astronomer, Edo Berger, who had spotted our names on the telescope schedule. The email implored us to change our observing plans and, instead of our planned targets, add our telescope's might to a hemisphere-wide hunt for a bizarre new object that would be kicking off as soon as it got dark enough to search.

Earlier that day, Edo had been in a committee meeting at Harvard when an alert had popped up on his phone. He was part of a special research collaboration that maintained an internal mailing list, ensuring he'd be informed immediately if this particular type of astronomical discovery ever happened. The email alert, which the collaboration had been anticipating for years, came in around midday East Coast time and pointed its readers toward a new discovery in the Southern Hemisphere. Edo and the other alert recipients had ten hours to swing their teams into action before it would be dark enough in Chile for the telescopes to open.

Even earlier that morning—at about 8:41 a.m. East Coast time— Cody Messick, a graduate student at Pennsylvania State University, had just messaged his astronomy research adviser to let him know he'd been working from home that day thanks to an injured neck. Cody was heading down the stairs at his home, ready for a quiet day of work, when an automatic text notification from his workplace popped up on his phone. The text was enough to freeze him on the stairs and blink at his phone as he digested the news.

An observatory in Washington State—one of only three observatories like it in the world—had just detected a gravitational wave.

o o o

Gravitational waves are best described as compressions in the fabric of spacetime. To understand what that means, imagine holding a Slinky between your hands. There are two different ways you can cause a wave in the Slinky. You could lift one end and watch a curved wave travel from one end to the other, but that's not quite what a gravitational wave is. If you instead held the Slinky and briefly pushed one hand closer to the other, making a compression wave that squeezes and stretches the coils as it travels, you'd have something similar to a gravitational wave. The only difference is that gravitational waves are moving at the speed of light and squeezing and stretching spacetime itself, along with everything in it; as a gravitational wave passes through Earth, the planet gets squeezed and stretched as well.

Gravitational waves fall into that wonderful realm of physics where things exist simply because the math of the universe insists they *should*.

Einstein's famous general theory of relativity, which describes the relationship between gravity, space, and time, is described by a series of ten
mathematical equations. In 1916, Einstein found that one consequence
of these equations was the existence of gravitational waves. The challenge
came from proving this (and, by extension, proving a vast swath of physics),
because while sheer math says gravitational waves can happen, it also says
they will be *minuscule*, a detail Einstein included in his original work. The
cataclysmic collision of two black holes with a combined mass sixty times
greater than our sun's will produce a squeeze in spacetime that's about one
thousand times smaller than a proton. Einstein himself believed these waves
would be too impossibly small to ever detect.

The solution ultimately came in the form of an interferometer, an
instrument that combines multiple sources of light. If this sounds familiar, that's because it is: radio astronomy uses interferometers to combine
data detected from multiple individual telescopes spread out over a well-
defined distance. Gravitational wave interferometers take the same basic
principle—combining light after it has traveled over a distance—to measure something different: exactly how far the light itself has traveled.

The basic principle is this: a gravitational wave detector can be built
out of two long, straight arms, arranged at right angles to each other and
connecting to a central building housing a powerful laser that gets beamed
down each arm. Perfect mirrors—known as "test masses"—can be attached
to the end of each arm to reflect the laser back and forth; together, the
whole system monitors the length of each arm with astonishing precision.
On a normal day, the lasers beamed down both arms will travel exactly
the same length, arrive back at the building at exactly the same time, and
neatly cancel each other out so that no signal is produced. However, a
gravitational wave passing through the planet will momentarily affect the
length of the two arms, squeezing one and stretching the other. When this
happens, the lengths of the arms temporarily differ very slightly, the test
masses move with them, and the reflected lasers detect this change and

produce a signal. An event like two black holes colliding produces a burst of gravitational waves and a distinctive signal in these detectors known as a "chirp." If we translate the frequency of the gravitational waves into sound, it resembles a single note lasting less than a second that swoops dramatically upward in pitch.

The concept is beautifully simple, but the execution is devilishly complex. The challenge is that a detector sensitive enough to detect the tiny perturbations produced by gravitational waves is also sensitive enough to detect, well, everything else. A well-built detector will surely pick up fluctuations in spacetime from colliding black holes billions of light-years away, but it will also be bombarded by the vibrations of earthquakes, trucks driving past, and hundreds of other things that can jostle the detector arms or test masses. The real challenge, then, isn't detecting gravitational waves; it's *not* detecting all the other extraneous signals—"noise"—that could be drowning out the tiny chirps announcing Einstein was correct.

For decades, three gravitational wave detectors worked diligently to become the world's most precise and sensitive astrophysics experiments, all operating on faith in physics: gravitational waves were real and would be detected, and it was just a matter of improving the instrument that theory said would detect them. Two detectors jointly formed LIGO, the Laser Interferometer Gravitational-Wave Observatory, built in quiet corners of the United States: LIGO Hanford in eastern Washington and LIGO Livingston in southeastern Louisiana. A third detector, the Virgo interferometer, was built in Santo Stefano a Macerata, a small village in Italy just southeast of Pisa, as part of a European collaboration. All three detectors began operating in the early 2000s, with thousands of scientists, engineers, and support staff constantly working to perfect every last detail of the facilities and become the first astronomical observatories in the world to measure not light but gravitational waves.

○ ○ ○

I headed out to visit LIGO Hanford on a quiet Tuesday in May, making the three-hour drive from Seattle and watching my surroundings morph from lush evergreen Pacific Northwest mountains to a color-leached open plain as I headed east. Washington's LIGO detector is situated in the Columbia Basin region of southeastern Washington, about ten miles north of Richland. After visiting mountaintop (and flying) observatories all over the world, I'd imagined a gravitational wave observatory as something vaguely dark and mysterious, a sort of eldritch corner of physics where researchers delved into the deep mysteries of gravity. Instead, as I approached LIGO itself, it reminded me strongly of SOFIA's home in Palmdale or the road leading up to the VLA; it's a similar feel of cutting-edge science being enabled by the sheer vastness of the open desert.

It was also almost funny to see how utterly, crashingly dull the place looked from the outside, considering the bleeding-edge science going on inside. A subtle sign marked the turnoff from the main road, an unremarkable

Aerial photo of the Laser Interferometer Gravitational-Wave Observatory (LIGO) in eastern Washington. *Courtesy Caltech/MIT/LIGO Laboratory.*

parking lot fed a couple of inoffensively blank buildings and a neatly curated visitor center, and upon close inspection, two bland gray concrete barriers covering the two arms of the interferometer extended off into the distance. Later, when I was taken up to the roof of the main building, I would see how ridiculously, *mathematically* straight these detector arms actually were. Each one stretches for almost two and a half miles, so far that they're built to take the curvature of the earth over that distance into account. Despite their length, only the first half of each arm is visible from the central building, with the second half completely obscured behind mid-stations built at the halfway point. Still, when you drive up to LIGO, they just look like…concrete.

The concrete barriers provide critical protection for the experiment. Each interferometer arm is actually a stainless steel tube about four feet in diameter, raised off the ground, and kept permanently at a near-perfect vacuum, even more pristine than the vacuum of space. The project had briefly considered just building these tubes in the open air, trying to keep costs down and worried about the expense of pouring miles of concrete arches, but eventually, the protective barriers were added. In the end, the concrete has saved LIGO more than once, shielding the vacuum tubes from multiple lightning strikes. Even with lightning rods, both the Washington and Louisiana detectors have taken hits that left burned gouges in the concrete. The Hanford site also survived a car crash. A driver illegally off-roading in a Jeep on a closed trail wound up crashing into the barrier hard enough to injure both himself and the Jeep, but the interferometer was unharmed.

The only problem the concrete barriers presented, albeit temporarily, was mouse pee. During the initial building phase, the barriers were filled with insulation, and after the work was completed, nobody saw a compelling reason to remove it. The insulation turned out to be a haven for mice looking for a handy spot to build their nests. Initially, nesting in the insulation was a nuisance but not a real concern, until the Louisiana site discovered a problem: the combination of mouse urine and the humidity of the area produced a bacteria that burned microscopic holes in the tubes and started threatening vacuum integrity. After this discovery, the mice were evicted, the insulation removed, and the holes fixed.

After arriving at LIGO Hanford, I was taken over to the main operations building and given a brief look at the control room (which looked like a scaled-down NASA mission control center with several workstations and a huge bank of screens across one wall) before gearing up to visit the laser and vacuum equipment area, or LVEA, a large central room housing the laser, some of the test masses, and other equipment. In my email conversations with the public outreach team at LIGO, they'd encouraged me to visit on a Tuesday specifically because the observatory would be shut down for its weekly maintenance, thus allowing me to visit parts of the detector like the LVEA that would normally be closed off. The reason wasn't *my* safety from walking near a powerful laser, though that was also treated very seriously; I received a pair of shiny reflective green goggles that would protect my eyes in case of a problem, along with a hairnet and shoe covers since the LVEA was a clean room. No, the concern was my actual footsteps; while the detector was running, anyone walking through the LVEA would create what they called "anthropogenic noise" simply by walking on the same concrete slab that the detector was built on. My guide explained that the LIGO crew would need to pull all of us out by noon, giving the reverberations from our apparently elephantine footsteps a little time to die down before switching things on again in the early afternoon.

It's one thing to *say* that LIGO can detect perturbations thousands of times smaller than a proton and quite another to see the practical requirements behind that sensitivity in action. The most spectacular technology is reserved for the test masses themselves and how they're suspended in the detector arms.

Each test mass is a circular mirror, just over thirteen inches across and almost eight inches thick, that weighs almost ninety pounds (bigger and heavier mirrors don't get shaken up as easily). The mirrors are made of

pure fused silica glass and designed to reflect the infrared light emitted by
the LIGO lasers. The test masses are suspended on a four-link pendulum
system (each link reduces the outside vibrations that are transmitted to the
mirror) and hung using glass fibers that are only four millimeters thick.
While each fiber can suspend nearly twenty-eight pounds of weight, they're
incredibly susceptible to other stresses: touching one with a finger can shatter the glass due to fractures caused by skin oils. The glass fibers, produced
on-site at Hanford, are ideal for suspending the test masses since there's less
friction introduced in the pendulum, and even the molecular motion within
the glass is an improvement over metal wires, reducing microscopic vibrations that are passed along to the mass.

I grasped the full extent of LIGO's sensitivity once I was back in the
control room and chatting with people about noise sources. Even with this
intensive attention to detail, monitoring and controlling for noise is still a
major element of LIGO's day-to-day operations. When a potential gravitational wave signal is detected in the rapidly processed data, one of the first
orders of business is meticulously tracking anything in the area that could
be contributing noise and causing a false alarm.

One pair of monitors on the right displayed sets of blue, yellow, and
red lines that slowly progressed across the screen, curving up or down with
time. I asked one of the LIGO employees in the control room what I was
seeing. They explained that they were monitoring noise sources: anthropogenic sources like footsteps, wind shaking the building, and so on.

I pointed at one screen, reading the label on the top. "What's 'wave-
driven microseismic noise'?"

"That's mainly ocean waves. Crashing against the North American
plate. It makes for some pretty steady background noise."

"You're *kidding* me." We were two hundred miles from the ocean.

On the same screen, there was a sudden and enormous spike in the
noise levels that had come in about twenty hours prior. I was visiting the site
less than a day after a 7.2-magnitude earthquake hit in Papua New Guinea;

from half a world away, the echoes from this earthquake were still enough to rock both the Washington and Louisiana detectors for hours, temporarily knocking out their ability to detect much of anything. LIGO has special procedures in place for earthquakes; the operators explained that since the waves didn't travel instantaneously but instead had to propagate through the planet, they could get anywhere from a few minutes to half an hour of warning when a substantial seismic disturbance was coming and respond accordingly. For small earthquakes, they would shift the test masses to a configuration that was less sensitive to gravitational waves but also less susceptible to getting knocked around. For big events, they would actually go into a kind of anti-lockdown mode and let the test masses swing free, taking them out of alignment but minimizing long-term effects from the quake until the tremors ceased and the detector could return to normal.

Sources of noise were seemingly never-ending. Early in LIGO's operation, the Louisiana site struggled to deal with noise from a logging site in the area. In Washington, they could pick up the annual spring dam release of the Columbia River. Other disturbances included the beating of helicopter blades flying high overhead, the thrum of propane engines from vehicles parked on-site, and rain. One recent and amusing source of noise in Washington had stemmed from the liquid nitrogen tanks used to cool the detector. In warm weather, ice would form on the pipes leading to the tanks, and enterprising ravens would start pecking at the ice as a handy source of water on a hot day. That *tap-tap-tap* was enough to launch a full-scale investigation into what was causing noise in the detector. The mystery was eventually solved when the scientists spotted gouges in the ice that were suspiciously well matched to the size of the ravens' beaks. The discovery was immortalized in a meticulous log entry (complete with photos of the peck marks, a raven caught in the act, and a graduate student mimicking the pecks to successfully reproduce the noise signal), the pipe leading to the tanks was altered so it would no longer accumulate ice, and the offender was profiled in an internal LIGO newsletter as "Thirsty the Raven."

Funnily enough, even for an observatory detecting gravitational waves rather than electromagnetic light, the conditions are slightly better at night: the cool air leads to lower winds, and the anthropomorphic noise drops as the number of truckers driving on the nearby highway decreases (you can then see the truck noise increase again in the mornings when drivers get back on the road, showing up first in the detector arm closest to the highway).

I was surprised to learn that just like at other observatories, LIGO employed teams of operators who worked in the control room and ran the detector 24/7 while the facility was observing. (I guess I'd imagined that someone just switched the laser from off to on and then patiently waited for signals to arrive.) Instead, the operators were constantly engaged, manually adjusting mirror positions to keep the alignment exact, watching for seismic alerts, controlling for as much noise as they could, and carefully keeping track of the noise sources that couldn't be prevented. In a striking parallel to Herman Olivares at Las Campanas Observatory, LIGO operator and graduate student Nutsinee Kijbunchoo is a cartoonist whose work often features the quirks of life as an astronomer working on gravitational waves.

o o o

Everyone who had even a passing familiarity with LIGO knew that discovering gravitational waves would be a huge accomplishment, an instant Nobel if ever there was one. This was not something that could be handled lightly. With a team of thousands spanning multiple continents, operating the human machine of running the facility and checking, analyzing, and publishing the results was at least as complicated as the massive machines themselves. In a stroke of brilliance, LIGO recognized fairly early on that its human components had to be tested for noise and sensitivity just as much as its interferometers.

This gave rise to the phenomenon of a "blind injection." Early on in LIGO's development, a tiny team was given a crucial but quiet task: they would insert fake signals, designed to look like real gravitational waves, into the stream of data coming from the detectors. This tactic is fairly common in research and serves as an important test of whether the people and software analyzing the data are capable of reliably detecting these sorts of signals. The LIGO blind injections, as it happened, wound up serving two additional purposes.

The first was a test of the full LIGO team. While the team knew that blind injections were possible, nobody apart from the handful of people who ran the blind injections knew which signals might be real and which were fake, so everyone was instructed to treat any signal they found as potentially real. This meant that they would analyze the data, crunch the numbers to determine what sort of astrophysical phenomenon had produced it (two black holes? how massive? how far away? and so on), and even draft a research paper announcing the momentous result, all without knowing whether it was a bona fide detection or a blind injection. The news of whether or not the signal was real would only be revealed at a final massive team-wide meeting, a process referred to as "opening the envelope." The nickname comes from the idea of someone opening a sealed envelope delivered by the blind injection team, Academy Awards–style, that reveals whether a fake signal has been injected into the LIGO data (in reality, the "envelope" has more recently been a flash drive containing a slide presentation). Then and only then would the team find out if they had been working on a blind injection or the real thing.

Blind injections also tested team members' ability to keep their mouths shut. A preliminary detection—the very first discovery of gravitational waves—was an immensely exciting result, but the data needed to be ruthlessly scrutinized for months to confirm it was a gravitational wave rather than a noise source, and it took hundreds of people just as long to unravel the underlying physics that would form the bulk of the groundbreaking research paper. LIGO needed to know that the news wouldn't leak, that nobody would blab to a friend or family member or colleague and risk letting word get into the wild before they were indisputably sure they'd made a detection. Carl Sagan popularized an aphorism that came to be known as the Sagan Standard: "Extraordinary claims require extraordinary evidence." LIGO wanted to be absolutely certain that no hint of this extraordinary discovery got out before the evidence was well and truly confirmed.

Word of the blind injections also spread to the broader astronomy community, a fact I'm quite sure was intentional. Even those of us who weren't at all affiliated with LIGO knew the stories of the sinister fake signals that could pop up and the fact that the team was largely in the dark. It provided a sort of funny two-sided insurance; if someone working with LIGO *had* spilled the beans, anyone who noticed would be hesitant to jump to conclusions, because who knew? Sure, their colleague might suddenly have a suspicious spring in their step while they went about their research, but they could just be hoodwinked by one of the blind injections.

The system worked wonderfully during earlier rounds of LIGO observations; occasional blind injections were detected, the team took the news (and the revelation that they'd spent months laboring away on a fake signal as a gigantic test) in good humor, and the world at large remained none the wiser. It seemed clear that when a real gravitational wave signal eventually arrived, the team would proceed precisely as it always had, with everyone (except for the small blind injection team) unaware of whether they had another fake signal in hand or the real thing.

The whole concept then proceeded to fail rather spectacularly when the first true gravitational wave detection happened on September 14, 2015. The LIGO team was just starting up the detectors again after an immense five-year upgrade to increase their sensitivity. That day, the detectors were on but not "officially" observing. Instead, they were in a phase known as an engineering run; the detectors were on and gathering data, but some auxiliary systems were still off-line. One of those systems was the blind injector.

Team members had been tinkering with it just the day before, trying to calibrate it properly, but it wasn't yet up and running.

This meant that when an immense chirp from two merging black holes arrived in the LIGO detectors early on the morning of September 14, several people immediately suspected that it might be the real deal.

The data was gorgeous, so much so that some members of the LIGO team who had been through previous blind injections were initially skeptical and thought it must be a handcrafted fake signal. Still, the few people in charge of blind injections quickly confirmed among themselves that the injector hadn't even been turned on. Several LIGO team members actually spent months exploring the possibility of a malicious injection, trying to figure out if someone could have sneakily gained access to the LIGO detectors and somehow faked a signal as part of an immensely elaborate scientific hoax. Eventually, this was also proven impossible. The gravitational wave signal LIGO had detected in September was spectacular, unambiguous, and *real*.

Whispers of some news from LIGO began to propagate out to other astronomers during those months, but this is where the large-scale insurance of the blind injections actually paid off. The collaboration was careful to keep the key detail that the blind injection system hadn't been working under wraps, so even those of us getting word that our LIGO colleagues seemed unnaturally excited about something couldn't be sure they weren't getting all worked up over nothing. In October 2015, I attended a conference of astronomers who studied dying stars, gravitational waves, and other time-domain phenomena, events that could rapidly appear and disappear in the night sky. I had signed up to attend a brainstorming session titled something like "Hey, what if we find gravitational waves one day?" The session attendees were about evenly split between classical astronomers and gravitational wave researchers. We all sat down, and one LIGO scientist almost immediately piped up and said, "Okay, who here is currently under press embargo?" The entire gravitational wave group, clustered along one side of a large conference table, raised their hands, failing miserably at containing mischievous grins. Eyes widened on the classical astronomer side of the room. "Okay, then," the LIGO researcher continued, settling into his chair with a satisfied smile. "Go ahead and start discussing. We're…just going to listen." The moment certainly suggested to all of us that *something* was happening in the gravitational wave community, but we were all quick to shrug it off, because hey, odds were that our colleagues on the LIGO side of the table had just been hoodwinked by another blind injection, right?

The gravitational wave detected on September 14, 2015, turned out to be the result of two black holes, twenty-nine and thirty-six times as massive as our sun, colliding and merging 1.4 billion light-years away. After months of research and checking and rechecking, a triumphant press conference was scheduled for February 11, 2016, to announce the discovery. Ironically, after years of practice and effort by the entire LIGO collaboration to keep the news secret until the moment of announcement, the press embargo was broken fifteen minutes early in what may be the most mundane way possible. Erin Lee Ryan, a research associate who was working at NASA's Goddard Space Flight Center, attended a party earlier that morning celebrating the impending announcement where NASA served a cake with "Here's to the first direct detection of gravitational waves!" iced on top. Erin enthusiastically snapped a photo of the cake and blithely posted it to Twitter, and science journalists everywhere pounced, thrilled to get a head start on breaking the official announcement. Decades of effort, multiple practice attempts from the blind injections, and the scrupulous silence of thousands of people had been broken by a sheet cake and a tweet. This is why I've never been too worried about my colleagues concealing the existence of aliens.

o o o

The announced discovery of gravitational waves was groundbreaking, making headlines worldwide and cementing a 2017 Nobel Prize in Physics for Rainer Weiss, Kip Thorne, and Barry Barish. Weiss and Thorne both made pioneering contributions to the theory and engineering of the detector, while Barish achieved the equally incredible feat of masterminding the scientific growth of LIGO from a forty-person group into a massive international collaboration. The discovery proved that the decades of determination and dedication and astounding engineering work had all paid off and ushered in a new era of astronomy.

The term *multimessenger astronomy* refers to the holy-grail concept of combining multiple types of data from a single object. In a field where we're largely restricted to working with the small amounts of electromagnetic light we can gather from distant objects, the science immediately gets more robust if we can detect some other quantifiable signal from the same object as well. We'd previously achieved this only once before, detecting both electromagnetic light and a smattering of neutrinos (tiny subatomic particles) from a very nearby supernova in 1987. The light and the particles together served as two "messengers" from space, giving us multiple tools for studying the supernova. Gravitational waves, as a third type of messenger, seemed set to usher in a new era of astronomy by serving as a completely new type of data that we could detect from cosmic events.

Still, we hadn't quite achieved the multimessenger designation with gravitational waves in 2015. The black hole collision LIGO observed was detected *only* in gravitational waves. As exciting as the discovery was, the scientific goalposts moved almost immediately, and everyone started looking forward to the next big discovery: an event that produced both a gravitational wave *and* a flash of electromagnetic light, a true multimessenger event.

Everyone recognized that this could be difficult. We'd detected black holes crashing into each other, but most astronomers agreed that they didn't expect to see any flash of light associated with these sorts of events.

The same shouldn't be true for colliding neutron stars. Neutron stars are the collapsed cores left behind after the supernova deaths of massive stars, the same objects that, when rapidly rotating, are sometimes detected as pulsars by radio telescopes. During collapse, the entire mass of the star's core gets packed into an area the size of a city, resulting in the unbelievably dense object that we know as a neutron star (a single teaspoon-sized scoop of neutron star would weigh more than a mountain). The neutron star only stops collapsing because of the Pauli exclusion principle of quantum physics, which states that subatomic particles like neutrons can't occupy the same quantum state within a system. If the neutron star tried to collapse further (and thus get even denser), the neutrons would be shoved together so powerfully that they and their neighbors would begin to violate this principle. To avoid this, the neutrons exert an outward pressure that can actually halt gravitational collapse. We're left with some of the weirdest objects in the universe: the tiny leftover husks of dead stars, supported by quantum physics and sometimes rotating hundreds or even thousands of times per second.

Neutron stars are extreme objects and close cousins to black holes, so when two neutron stars in a binary system spiral into one another and collide, it's an extremely gravitationally violent event. The collision and merger of two neutron stars should produce a gravitational wave, similar to the chirp of two merging black holes but with a longer duration and lower energy. Crucially, a neutron star merger is also expected to produce a brief blast of high-energy electromagnetic light, detected on Earth as a burst of gamma rays lasting less than two seconds, and a much longer flash of light known as a kilonova. Weaker than a supernova, a kilonova nevertheless acts like a giant signal fire after a neutron star merger, blazing brightly and then burning out after several days. This means that if a gravitational wave signature from two merging neutron stars was ever detected, electromagnetic observatories all over the world would have to be poised to commence an immediate, frantic, and worldwide search for the light from the kilonova before it faded away.

The search is much harder than it sounds. Gravitational waves are extremely difficult to pinpoint in the night sky. The LIGO detectors in Washington and Louisiana can, together, estimate roughly where a gravitational wave is coming from based on which detector picks up the event first and a few details of the signal itself. Combined with the Virgo detector in Italy, the three observatories can do a better job of triangulating where a signal is coming from, but they still narrow it down to a sizeable swath of sky that would need to be searched for any possible light that could be attributed to colliding neutron stars. A team of kilonova hunters would need to prove that the object they identified as the kilonova had only appeared after the gravitational wave event, meaning they would need to be lucky enough to have prior observations of that patch of sky available for comparison. They would also need to prove it was indeed a kilonova rather than some other transient event, meaning they would need to closely observe the object to prove that it matched theoretical predictions. This requires the ability to work both very fast and very carefully and to utilize huge libraries of data to find those pre-event comparison observations.

As the gravitational wave observatories of the world continued to work, detecting more bona fide gravitational wave signals from merging black holes, teams of light-based astronomers were poised like arrows in drawnback bow strings. Different research groups had formulated different techniques for searching large swaths of sky, called dibs on different telescopes or claimed priority at following up different types of candidate signals, and established detailed plans for exactly how they would spring into action when a signal from merging neutron stars was eventually detected. This was the very type of signal that arrived early on the morning of August 17, 2017.

o o o

When Cody Messick was standing frozen on his stairs, gaping at his phone, he wasn't entirely convinced that he was seeing a real gravitational wave. The signal described in the automatic text alert was certainly unusual—it had only a one in ten thousand likelihood of appearing in the data by chance—but it didn't look like the previous binary black hole mergers. It was longer, weaker, and had only appeared in the LIGO Hanford detector in Washington. Team policy had long been to ignore signals unless they'd been picked up by both detectors, since a one-detector event was far more likely to be local noise than a real signal from space. Cody, however, was one of the team members who had specifically set up an alert for himself so he could investigate single-detector events on the off chance that one might eventually be real. Intrigued by the low false alarm rate, he sent it off to his adviser, Chad Hanna, who recognized the signal as similar to the prediction for a neutron star merger. He also noticed that the Fermi Gamma-ray Space Telescope had, 1.7 seconds after the gravitational wave signal, detected a two-second burst of gamma rays, precisely the sort of high-energy signature expected as the first electromagnetic counterpart to a neutron star merger.

Cody and Chad, along with a few other team members, immediately jumped onto phone calls and team chats to begin digging into the data. They quickly confirmed that there were no contaminants and no other noise sources in the data. The practice of blind injections had long since ceased, and the simultaneous gamma-ray burst that had been detected was an incredible smoking gun. Chad agreed he would let the larger LIGO collaboration know, then sent a follow-up message a few moments later explaining he was shaking too much with excitement to type up the email. In the end, Cody sent the first email to the collaboration: they had a candidate signal that looked like a binary neutron star merger, it was almost certainly not a false alarm, and it had coincided with a gamma-ray burst.

The email prompted a virtual stampede of team members piling into the chat and discussion to start unraveling the data. (Nutsinee, the cartoonist LIGO operator, drew an illustration of this sort of discovery that was particularly apt, depicting a groggy operator waking up in bed, picking up their phone, and being practically blasted off their mattress by the onslaught of messages clamoring about the detection.) The first big question the team needed to address was why only LIGO Hanford had reported a detection. The Virgo detector in Italy had been having a data transmission problem, but if it was a real gravitational wave detection from deep space, then LIGO Livingston in Louisiana should have spotted the signal and sent out an alert as well. What had gone wrong?

The answer was immediately obvious when someone pulled up the LIGO Livingston data from that time frame. In the data, a human eye could immediately see the slow, swooping chirp signal of a binary neutron star merger, but slapped on top of it was an ugly blast of detector noise known as a glitch, like a photographer's thumb positioned awkwardly in the corner of a snapshot. Since the computer had been taught to never send out an alert if the data was contaminated by a glitch, the LIGO Livingston detection had gone unannounced. Fortunately, the glitch could be measured and removed, and the team was left with a pair of beautiful binary neutron star merger signals from both detectors, detected three milliseconds apart and in the Louisiana detector first.

Once the Louisiana data was added to the puzzle, the final bits of guarded scientific caution fell away, and the team went wild with excitement. With a gravitational wave *and* a gamma-ray burst, this new event, dubbed GW 170817 to indicate a gravitational wave detected in 2017 on August 17, was the first candidate for a multimessenger detection involving gravitational waves and electromagnetic light.

Still, the gamma-ray data was a thoroughly encouraging sign but not a slam dunk. After all, it could have just been random chance, the burst had been incredibly brief, and gamma-ray bursts are also difficult to pin down location-wise. Nobody could say anything for sure unless astronomers were able to find a clear, bright new flash, captured unambiguously with a ground-based telescope, that could be attributed to the event's kilonova. With the trifecta of a kilonova, gamma-ray burst, and gravitational wave detection, then and only then would astronomers finally be certain of what they had found.

With data from the Virgo detector quickly added to the mix, the gravitational wave teams were able to significantly narrow down the part of the sky where the signal had come from. The binary neutron star merger had happened somewhere in the Southern Hemisphere and somewhere within an area that was the size of about 150 full moons. Still, this left an enormous area of the sky to search. Knowing this, LIGO send out a broader announcement to a list of astronomers who were specifically interested in chasing gravitational waves, all of whom worked in teams that were permanently ready to pounce when an event like this happened.

This was the announcement that prompted Edo Berger's email to Phil and me later that afternoon and that threw a large swath of the dying star and gravitational wave communities into a frenzy as they prepared to commandeer the telescopes of the Southern Hemisphere and search for any sign of the kilonova associated with GW 170817. In the end, more than seventy telescopes were involved in the observations.

Once darkness finally fell in Chile and observations began, it took less

Cartoon by Nutsinee Kijbunchoo, depicting news of a binary neutron star merger detection. *Credit © Nutsinee Kijbunchoo.*

than two hours for teams to start hitting pay dirt. The kilonova itself was almost anticlimactic in its simplicity when it was found: a little blue dot, sitting in the outskirts of a wholly unremarkable-looking galaxy about 130 million light-years away, that most definitely hadn't been there before. It was a perfect match to the distance determined from the LIGO data and to the predicted appearance of a kilonova.

Little blue dot or not, it's almost impossible to overestimate how deliriously excited astronomers can get over a tiny speck of light in the right place, and as different groups independently spotted this new source, it prompted a range of reactions. Ryan Chornock, a member of Edo Berger's team, had been poring through their data to spot any new flash that could come from a kilonova and circulated an email to the whole team that read, in its entirety, "holy shit," with the discovery image attached. Charlie Kilpatrick, an astronomer working with another team, opted for the understated comment "found something" in a group chat, followed shortly by a screenshot showing the kilonova. In total, five teams independently discovered the kilonova counterpart to GW 170817 within about twenty minutes of one another.

The discovery was just step one. Once the kilonova had been found, astronomers immediately set about wringing every bit of science they possibly could out of this one patch of sky and speck of light. Imaging and spectra, X-ray and ultraviolet and optical and infrared and radio light, at small and large observatories, observing the kilonova and its surrounding galaxy... almost every telescope that could point at this patch of sky did, plunging astronomers who studied gravitational waves and astronomers who just happened to be at the right telescope at the right time into the same all-hands-on-deck race.

To complicate matters even further, a decent fraction of these people were headed out to some of the most remote corners of the United States for the August 21 solar eclipse. Mansi Kasliwal, a team leader for one of the follow-up groups, had volunteered for a massive public eclipse event at

Caltech and found herself juggling eclipse observations, ten thousand eager participants, and her own toddler while coordinating news from her entire team and time-sensitive phone calls from the Hubble Space Telescope asking for final observing plans. On another team, Maria Drout and several other astronomers had volunteered for an eclipse event at a school in Idaho. Months before the eclipse, they had planned a driving and camping trip through Utah, but everyone in the car wound up involved in the gravitational wave follow-up. Maria and her colleagues ended up tethering laptops to tenuous cell phone signals in the back seat of the car, stopping at restaurants with Wi-Fi to download new data, and analyzing the data in a tent.

GW 170817 was also yet another case where any attempts at secrecy failed spectacularly. Rumors of the discovery hit the internet almost instantly, with the true nail in the coffin coming from a Twitter account belonging to the Hubble Space Telescope. About a year earlier, the account @spacetelelive had begun posting simple automated tweets reporting what Hubble was observing—"I am looking at X object with Y camera for Dr. Z" or similar—drawing from the Hubble database of targets and observing plans. Immediately after spotting the kilonova, Edo Berger's team submitted a frantic request to Hubble, begging them to point at the binary neutron star merger to observe the fading ultraviolet light that couldn't be detected from the ground. In their urgency, they gave their proposed target the simple and straightforward name "BNS merger," an abbreviation that any astronomer or science journalist would instantly recognize. The team recognized their mistake before the observations began and called Hubble to edit the target name, but the change never went through. Hubble pointed to the kilonova, took their proposed observations, merrily tweeted out that it was observing a BNS merger, and the cat was out of the bag. Articles speculating that we'd detected the first multimessenger gravitational wave appeared online within hours. Andy Howell, on yet another team scrambling to observe the counterpart, tweeted a bit more obliquely on the first night of the follow-up observations: "Tonight is one of those nights where watching the

astronomical observations roll in is better than any story any human has ever told."[29]

The discovery of the kilonova and the resulting sprint for follow-up observations also highlighted a fact that scientists everywhere tend to try and politely ignore: politics. There *was* science behind the speed, of course—the kilonova was fading with every passing moment—but there was also the more basic human motivation that everyone wanted to be first. Within the tiny astronomy community, most teams knew exactly who they were racing. Some of those teams already hosted rivalries that went back years, and others had previously maintained friendships that quickly soured as they fought to get first dibs on different powerful telescopes. As with any group of people, some astronomers were purely opportunistic and eager to grab for the fame and recognition that would come with being front and center in the inevitable press coverage, while others spurned the whole mess completely to try and focus on the science. Most people fell somewhere in the middle, trying to work as carefully and as quickly as possible while keeping the careers and dreams of their team members—particularly more junior scientists such as graduate students and postdoctoral researchers—in mind. In the end, the follow-up circus quickly devolved into an unholy mess of sniping, rescinded handshake deals, and backroom squabbling, something that is still greeted today with a mix of embarrassment and disappointment in the community. For LIGO's part, they mostly looked on in bemusement. Their several-thousand-person team ran much more like a huge and disciplined physics operation, so the spectacle of tiny clusters of astronomers running around like chickens with their heads cut off largely prompted them to back off and stay squarely on the gravitational wave side of things.

Political battles aside, GW 170817 ended up producing some beautiful science. Despite the initial scramble, the data was eventually compiled into a hefty series of peer-reviewed papers and announced (officially, at least) only after it had been properly scrutinized. The main paper from LIGO announcing the gravitational wave detection from a binary neutron star merger had 3,684 authors, and the *Astrophysical Journal* pulled out all the stops to quickly and carefully peer review research papers on the topic as they came in and release a special issue devoted solely to thirty-three papers from the full community of follow-up teams that covered every last detail of the groundbreaking event.

o o o

In the aftermath of the first few gravitational wave discoveries, LIGO has now flipped from a seemingly long-shot physics experiment to a glorious triumph of engineering and perseverance. With over a dozen gravitational wave detections in hand, the secrecy around these signals has been completely dismantled; LIGO now announces new detections on Twitter once they've been verified. Astronomers have moved on from their wild race to find the first binary neutron star counterpart. Recent conferences have included discussions of coordinated and collaborative follow-up for similar future events that everyone is certain will arrive sooner or later. While future observations will likely be a bit smoother, it's clear that the scramble to study the GW 170817 kilonova wasn't the last of its kind.

Nor was it the first. The gravitational wave element was certainly brand new, but the art of identifying fleeting changes in the sky and chasing them down before they disappear again is a long and time-honored field in astronomy. It has, in the past few decades, also begun driving a significant shift in how we use telescopes.

CHAPTER ELEVEN

TARGET OF OPPORTUNITY

Oscar Duhalde can lay claim to a truly unique astronomical discovery. He is the only person on the entire planet—and probably one of only a handful of people in the history of the human species—who has discovered a supernova with the naked eye.

Oscar is a telescope operator at Las Campanas Observatory in Chile. Early on the morning of February 24, 1987, he was working at the mountain's one-meter, guiding the telescope by hand as the two astronomers observing that night took exposures with a CCD camera. Hand guiding is a straightforward but fairly tiring process, with the operator constantly tweaking the telescope's position to make sure the object the astronomers are studying stays precisely positioned in the center of the telescope's field of view. Oscar, after over four straight hours of guiding that evening, finally let the astronomers take over so he could take a brief break around 2:00 a.m. He headed downstairs from the control room and prepared some coffee, and while it brewed, he stepped outside to admire the sky.

When he looked up, he noticed that something was slightly different.

Overhead was the Large Magellanic Cloud, nicknamed the LMC, a small satellite galaxy of our Milky Way about 163,000 light-years away. At that distance, the miasma of individual stars in the galaxy melts into a sort of bright fog rather than a visible collection of stars, giving it its apt name. A practiced eye can spot a few familiar features—knots of newborn stars, bright stellar clusters, or patches of light-blocking interstellar dust—but even for most astronomers, the LMC is a lovely feature of the southern sky rather than something they can map from memory.

As it happened, though, Oscar knew the LMC like the back of his hand. One of his early positions at the observatory had been serving as the night assistant to an astronomer named Allan Sandage, back in the days of photographic plates. A giant of observational astronomy whose legacy stretches back to the 1950s, Allan had spent a great deal of time observing the LMC. He had taken what must have been hundreds of photographic plates during his time at Las Campanas, and Oscar, as his night assistant, had *developed* hundreds of those photographic plates. He knew every last detail of the LMC.

On this night, there was an extra star in it.

Oscar stared for a little while, surprised to see an odd—and *bright*—new star in a galaxy that hadn't changed in all the years he'd been working at telescopes. Pondering what this new star could be and why he hadn't seen it before, he headed indoors to check on his coffee but kept being drawn back outside. At first, he thought it might be a satellite, but the star remained bright and stationary. Another coffee check and another peek outside confirmed that it was, undeniably, still there. *What the hell is that?* he wondered.

In the back of his mind, Oscar faintly recalled some research groups that had recently been searching for supernovae, and he knew that these distant stellar explosions could appear as bright new points of light in nearby galaxies. However, at the time, most searches were focused on enormous galaxies, filled with heaps of stars that might be ready to die, rather than shrimpy little galaxies like the LMC. Still, it planted the first germ of an idea in his head that the odd new star he'd just spotted could be something interesting, and he made a mental note to talk to the observers when he headed upstairs.

Unfortunately, Oscar arrived back in the control room to a beeping computer and two astronomers who were happy to have their operator back and ready to move to a new target. He hurried to point the telescope, turn the dome, and prepare for the next observation, momentarily forgetting all about his strange new object.

About two hours later, Ian Shelton, an astronomer who had been observing at another telescope on the mountain, burst into the control room and hurried past Oscar, who kept working as Ian started chatting excitedly with the other astronomers in the room. Oscar overheard bits of the conversation: Ian had been observing photographic plates at his telescope and, after closing early due to high winds, had developed the plates and compared them to images from the night before. With two images of the LMC sitting practically side by side, the appearance of an extra star had jumped out at him. Could there really be some strange new object in the LMC?

Across the room, Oscar looked up. "Ah yes! I saw it when I went out."[30]

The frantic burst of activity that followed over the next few days would soon name the odd new star Oscar had first spotted in the LMC SN 1987A, the first supernova discovered in 1987. The supernova was bright enough that once they knew where to look, Ian and the other astronomers were easily able to spot it. Later records showed that other Southern Hemisphere telescopes in New Zealand, Australia, and South Africa also observed the supernova later that night, but by all accounts of the timing, Oscar's naked-eye observation was the first.

Astronomers had been discovering supernovae for years, but they had all been much farther away in relatively distant galaxies. The last one visible from Earth with the naked eye had been a cool 383 years earlier, in 1604, several years before the telescope was even invented. A supernova that was practically in our backyard prompted a mad scramble to study the burst of light before it faded away as the stellar explosion dimmed. SN 1987A also became the first instance of multimessenger astronomy, producing neutrinos detected by experiments in Japan, Russia, and the United States. Today, SN 1987A remains the closest supernova that we've seen in the modern era, and it is still among the most dramatic examples of what we call "target of opportunity" astronomy.

o o o

It's easy to think of the sky as static and unchanging. Astronomical time usually unfolds over millions or billions of years. When we look up from night to night, the sky largely looks the same to us. The moon waxes and wanes, the planets of our solar system move across the sky, and we see different parts of the celestial sphere depending on the season, but the stars and constellations themselves appear to stay the same.

The surprise, then, is the things changing on a timescale that's short even for humans: days, hours, even seconds. For all that they've been around for millions or billions of years, the death of a star, a flare from an existing star, or a flyby from an asteroid or comet can sometimes happen surprisingly fast.

A supernova is the explosive end result of a dying star. Stars fight back against the inexorable inward crush of gravity by essentially operating fusion reactors in their cores, fusing hydrogen into helium or helium into carbon as a source of energy. The most massive stars in the universe go tearing through different fuel sources near the ends of their lives, desperately fusing oxygen and neon and silicon in a doomed attempt to stay alive for even a few more days. When this finally fails, the star ends up with a core of iron, which *takes* energy rather than *produces* energy when fused. At this point, gravity finally wins the battle it's been waging for millions of years, and the star's core implodes in less than a second. The outer layers go tumbling after it before bouncing off the collapsed core remnant and blasting out into interstellar space at speeds of seventy million miles per hour. The immense blast of light from this rebound explosion outshines the entire galaxy that the star has been living in.

You'd think a fireworks show of this magnitude would be pretty hard to miss.

In reality, it's shocking how difficult it is to spot an exploding star. Part of this is the simple issue of distance—"brighter than an entire galaxy" is impressive, but even those galaxies require sizable telescopes if we want a good look at them. Every supernova we've ever studied with modern telescopes—even the one that Oscar discovered by eye—has been in another galaxy, and for decades, finding them depended on the hard work of devoted amateurs and doggedly persistent supernova hunters who would image nearby galaxies over and over, searching for the sudden appearance of what would look like an especially bright new star (the very name *supernova* derives in part from the Latin word *novus*, meaning new). The average supernova brightens and then dims again in a matter of days, so if we don't catch a glimpse of one in the week or two when it's near peak brightness, the opportunity to find it and observe it is permanently lost.

This is the challenge of studying exploding stars—they are there one day and gone the next, leaving astronomers scrambling to try to explain what they saw based on whatever paltry bits of data they were able to capture in time. The need to respond quickly has spawned a new class of observations known as target of opportunity, or ToO. With a ToO, an observer can essentially propose to commandeer a telescope whenever a particular type of explosion is detected, jumping in from afar to quickly point at the new discovery (and sidelining whoever was originally scheduled at the telescope). A prompt response has become the holy grail of supernova ToO observers. Capturing the first hours or even the first minutes of a supernova can offer a glimpse of the earliest moments after death, literally illuminating the material closest to the star and telling you a great deal about the strength and speed of the explosion and what manner of extreme physics might be propelling it into space.

The only problem comes when someone triggers a ToO observation over a false alarm.

Brian Schmidt, who won the 2011 Nobel Prize in Physics for his team's groundbreaking work on using observations of supernovae to study the expansion of the universe, sent a breathless email around to a large research group one evening. It was a particularly dark and clear night, and he'd spotted what looked like an unfamiliar new star in the constellation Scorpio, near the horizon. Scorpio is a fairly bright and familiar constellation, so the appearance of a surprise star was cause for considerable excitement. Brian's email went out to well over two hundred people, excitedly reporting that he had spotted a bright new object! Visible with the naked eye! It was in Scorpio, right on the horizon! A local amateur had even confirmed it! The unspoken message, of course, was that everyone should spring into action to start observing this mysterious new object immediately, in the hopes that it might be the next naked-eye supernova and, even better, a supernova happening in our own galaxy.

A Milky Way supernova hasn't been seen since the year 1604. Based on the quantity and ages of stars in the Milky Way, our current guess is that we should see one of our stellar neighbors exploding about once every hundred years or so. By those numbers, the next one is due any day now.

At the positively next-door distances of our own galaxy, the appearances of Milky Way supernovae can be *extreme*, far more dramatic than the little extragalactic blips that we currently study. On July 4, 1054, the supernova death of a star only 6,500 light-years away grew so bright that it outshone every other object in the sky besides the sun and the moon. It was visible in the daytime sky for two weeks and was immortalized in Chinese, Japanese, and Arabic historical records and in an Ancestral Puebloan pictograph in Chaco Canyon, New Mexico. The remnant of that supernova, the Crab Nebula, is one of the most famous and well-photographed objects in today's sky.

A Milky Way supernova in this day and age would be nothing short of spectacular. I've always found it fun to think about what might happen if a nearby star exploded tomorrow. The initial sighting—a sudden point of light appearing in the sky and quickly getting brighter and brighter until you could read by it at night and spot it during the day—could risk prompting some real panic depending on the current geopolitical situation. Once it had been successfully identified as a supernova, a worldwide frenzy would commence, centered around stellar astronomy—this thing, after all, would be unmissable for at least half the planet. It would lead the news. It would get its own hashtag and start trending on Twitter. Nighttime talk show hosts would make jokes about it. A photo of it would wind up on every smartphone in the hemisphere. And the observational astronomers of the planet, myself included, would go absolutely out of our minds with excitement.

Brian Schmidt knew this and also knew that in this situation, being the first and fastest observer to point a telescope at the new supernova meant everything, both for the science and for securing a front seat to the ensuing mayhem. As his team swung into action, Brian began placing calls of his own, starting to mentally puzzle out who could contact which telescope for a ToO observation to start pursuing what would surely be the supernova of the century, and dug through object lists to determine what newly dead star might have been at that spot in the sky.

About thirty minutes later, the urgency dropped with his follow-up email: "Never mind. Am a horse's ass. Mercury is at that position."

o o o

Astronomy, like any other field, loves its false alarm stories: the microwave ovens mistaken for radio bursts, the planets misidentified as dying stars. They serve as excellent hands-on proofs of the scientific method for young observers, teaching us how to be skeptical and how to "think horses, not zebras" when hoofbeats show up in our data. In a field of science where we can convincingly prove that two colliding stars, supported by principles of quantum physics, can produce flashes of gamma rays and compression waves in the fabric of spacetime, it's important to keep our sense of skepticism well calibrated.

Back in 2005, during the same summer internship that had me giving tours at the VLA in New Mexico, I had been assigned a project working with some data from the Westerbork Synthesis Radio Telescope in the Netherlands, an interferometer with fourteen radio telescopes arranged in a straight line. In the midst of scrolling through image after image of rather dull data, I was thrilled to suddenly spot a new and unexpected signal popping up in several of the files. I took careful notes on what I had found, as it seemed to be varying—the signal was definitely brighter in some data than in others. I looked through old research papers and saw no mention of a similar signal. I started shoving plots and graphs under the nose of anyone I could find to see what they thought of this fascinating new discovery. Surely, I'd uncovered something exciting! Nobody had ever reported a signal like this before! It was changing rapidly with time! I was working just up the road (in New Mexico terms) from the VLA where *Contact* had been filmed; I'm willing to admit that the word *aliens!* drifted through my brain for half a second.

It took less than a day for the bubble to burst. When I sorted the data out between the fourteen different radio telescopes, it quickly became clear that my varying signal was always strongest during business hours and in the dishes closest to the observatory's administrative building. This was years before Emily Petroff and her research team would explain the mystery of perytons, but I knew that anything extraterrestrial in origin should have been uniform across every dish. I wasn't detecting aliens; I was detecting someone getting a fax or heating up a stroopwafel. Ah, well.

Still, strange signals are enticing to chase, if for no other reason than it's always exciting to find something *truly* weird. The odd, the surprising, and the unexplained are all fertile ground for groundbreaking discoveries.

In May 1962, Daniel Barbier and Nina Morguleff were observing at the Haute-Provence Observatory 193-centimeter telescope in France, cycling through a target list of nearby stars and taking spectra to analyze the chemical compositions of the stars' atmospheres. Like most observers, they were extremely familiar with the data they were working with after months of research and could, at a glance, identify the telltale signatures of common elements by spotting spikes or dips in their spectroscopic data that corresponded to specific wavelengths or colors.

Usually, the chemistry of a star stays extremely stable from moment to moment, so the French observers were surprised when they analyzed three observations of the same star and spotted bright orange light from the element potassium in one spectrum. The potassium itself wasn't too out of the ordinary, but its sudden appearance in just one observation out of three was certainly odd and suggested that they might have seen a new and strange type of stellar flare.

Stars flare all the time, a possible consequence of magnetic energy built up in their outer layers and then released. Our own sun regularly throws out small solar flares, spitting light and plasma and charged particles. That said, flares large enough to see from halfway across the galaxy are a good deal more dramatic. Catching one is a lucky break—typical stellar flares only last for a few minutes—and studying them is a valuable way to learn more about the star's inner physics, outer layers, and even how flares might impact the possible presence of life on any nearby planets. A new type of flare suggested possible signs of a new type of stellar physics.

Daniel and Nina excitedly wrote up a short summary on this newly discovered "potassium flare" for publication in the *Astrophysical Journal*. The excitement grew further when two more "potassium-flare stars" were discovered in the next few years. In astronomy, where data is so often limited by what we're lucky enough to see, one object is an oddity, but three is practically a category. By 1966, potassium-flare stars appeared well on their way to becoming a legitimate new discovery.

There was just one problem: potassium flares had only ever been detected at Haute-Provence Observatory, and the three flaring stars had all been discovered by the same collaboration of astronomers. This was, in fact, the *only* thing the stars had in common. One was sun-like, one was hotter, one had an odd magnetic field, but none of them shared any common ground that might explain a sudden blast of potassium.

Bob Wing, Manuel Peimbert, and Hyron Spinrad, a group of astronomers in California, were extremely interested in the prospect of true potassium-flare stars but also skeptical of the Haute-Provence discovery. After all, they had surveyed 162 stars themselves at Lick Observatory in search of potassium flares and not caught a single one. It begged the question: What else could the telescope possibly be seeing that would produce a brief flare of potassium emission?

The source turned out to be a bit more local than expected. Some of the French observers and technicians were smokers; Daniel Barbier in particular was known for smoking a pipe during observing runs.

Potassium, as it turns out, is the strongest feature in the spectrum of a match.

The California group investigated this during an unusual observing run at the Lick Observatory 120-inch telescope, standing at various positions near the spectrograph and lighting matches to see if they could produce bespoke potassium flares in their data. They also contacted Yvette Andrillat, one of the French observers who had observed potassium-flare stars, to explain their theory. Like most scientists, astronomers love a good puzzle, even if it proves their own research wrong in the process. Yvette immediately carried out her own similar experiment at Haute-Provence. As it turned out, the room where the French spectrograph was kept (apparently also a convenient room for a midnight smoke break) used a rotatable glass plate as part of the instrument that could reflect light from the match strikes and send it directly into the spectrograph detector.

The results of the whole endeavor were presented in one of the more delightful astronomy papers ever written. Credit is given to George Preston, who had gamely approved a telescope proposal that must have essentially boiled down to "Hey, we'd like to run around the telescope playing with matches." The whole experiment, absurd or not, was meticulously documented: the group was careful to test book matches, kitchen matches, and safety matches and noted from their communication with "Mme. Andrillat" that "There appear to be no significant differences between French and American matches."[31] The astronomy community got a handy published table listing the key elements present in the spectrum of a match, the Haute-Provence spectrograph room was declared nonsmoking, and the mystery was solved.

Still, the matches made one last appearance. In 1958, George Wallerstein (my University of Washington colleague who recently celebrated sixty years of observing) had observed potassium emission in the spectrum of an unusual red supergiant star named VY Canis Majoris. At the time, he had attributed it to the rare physical conditions present in the outer layers of the star. Almost a decade later, he happened to be the reviewer assigned to read the Wing, Peimbert, and Spinrad paper. A couple of years later, he published a new paper on VY Canis Majoris with his collaborators, calmly noting that *real* potassium emission in this star had been confirmed by new observations and that the match strike explanation "does not apply since the observer does not smoke."[32]

o o o

Target of opportunity astronomy grew out of the urgency that comes when we think we've spotted something new and need to get data on it as quickly as possible before it disappears. Supernova explosions and (real) flaring stars are just two examples of what sometimes gets referred to as "time-domain" astronomy, the practice of studying rapidly varying things in the night sky. Some time-domain subjects, like exploding stars and flaring stars, are fleeting events that must be pounced on and studied before they disappear. Others, like newfound asteroids scooting across the sky or stars that vary regularly as a function of time, need to be tracked at regular and sometimes inflexible intervals.

When SN 1987A was discovered, supernovae were announced via something called the Central Bureau for Astronomical Telegrams. Today, announcements like this are posted online to alert people to new supernovae, other transient phenomena, and the odd false alarm: every once in a while, an enthusiastic astronomer will make a similar error to Brian's, mistaking a planet for a brand-new supernova. In 2018, an astronomer excitedly posted on the Astronomer's Telegram website to report a "very bright" new object that had appeared in the constellation Sagittarius. Forty minutes later, he sheepishly circulated an update: the bright object was simply Mars, passing through Sagittarius as it followed its usual orbit around the sun. The community understood both the mistake and the enthusiasm, and the whole thing was taken in stride, but not before the administrators of the Astronomer's Telegram website drew up a certificate for the astronomer congratulating him on discovering Mars.

Announcing a find is only half the battle; the real challenge then becomes getting access to a telescope for a ToO observation. Usually, time on a telescope is apportioned well in advance, and months go by between writing a successful proposal and actually sitting down at the telescope to get the data. ToO observations, by contrast, have to happen within hours or even minutes, and how they're done depends very much on the astronomer, the target, and the observatory we're hoping to use.

Sometimes, in a handy stroke of luck, the observer is actually *at* the telescope, either by happenstance or as part of a concerted effort to observe exactly these sorts of moving and changing objects. In 1992, Dave Jewitt and Jane Luu were observing at the University of Hawaii eighty-eight-inch telescope and undertaking a dedicated search for Kuiper Belt objects. The Kuiper Belt (named for airborne astronomy pioneer Gerard Kuiper) is a wide ring of small solar system objects, made mainly of rock and ice, that begins just past the orbit of Neptune and stretches to a distance of 4.6 billion miles away from the sun. Today, Pluto and its moon, Charon, are considered members of the Kuiper Belt, but in 1992, with Pluto still designated as a planet, Dave and Jane were leading the search to discover the first object in this hypothesized fleet of small solar system objects.

Their method employed the same technique as many other searches for moving or changing objects in the night sky, blinking between two images taken of the same region of sky to search for anything that had changed. In Dave and Jane's case, they were taking a series of four images at each spot they observed in the sky and blinking between the images to search for objects that moved. They were particularly interested in objects that were moving *slowly*. Speedy small objects could be nearby asteroids, but slow-moving objects were likely farther away. (This is the same principle that you notice when watching scenery zip past from a moving car or train. Relative to you, nearby trees or buildings are flashing past quickly, while more distant landmarks appear to move at a relative crawl.)

They had been searching for five years when, on one August evening in 1992, they began blinking between the first two images they'd observed in a new patch of sky and noticed a very-slow-moving object that, to their cautious excitement, looked exactly like what they'd expect for a Kuiper Belt object. The third and then fourth image confirmed their expectations, with the object moving in a gloriously slow and straight line. At this point, they were, according to their program, supposed to move on to a new spot in the sky, but instead, they stayed put, tracking this strange new object for the rest of the night and gathering as much data on it as possible. The change in plans was partly done for the science but also partly done out of caution; the thing was moving, after all, and what if they moved on and then never found it again? In the end, they were able to measure a distance and a size for the object, confirming it as the first discovery of a Kuiper Belt object. Today, it's known as Albion, a rocky body over seventy miles in diameter orbiting the sun from just over four billion miles away, and it's one of an estimated thirty-five thousand objects like it in the Kuiper Belt.

In other cases, astronomers will rely on friends and acquaintances or simply the power of persuasion to capture and follow an exciting new object in the sky. Edo Berger's email to my colleagues and me, asking us to observe the GW 170817 counterpart, was one of these; he saw our names on the telescope's schedule and knew us well enough to send an email and ask if we wouldn't mind interrupting our time for his urgent observations. In other cases, astronomers can simply call or message telescopes, eventually reaching the observers themselves and asking if they wouldn't mind adding a target to their night or briefly interrupting their scheduled plans.

This is a fairly convenient approach, catching someone who's already at a telescope and ready to observe. That said, it also relies on the opinion of the observer. Astronomers are well within their rights to refuse such requests, and while people are often happy to slightly rearrange their evenings for an exciting new discovery, others sometimes can't disturb their observing plan or would simply prefer not to. (I know more than one astronomer who has simply responded to these calls with a curt "wrong number" and a hang-up, happy to stick with their long-planned science rather than upending their schedule for yet another exploding star.)

This approach can also turn into a bit of a race, with multiple groups competing to be the first to stake a claim on an observer's time. During one of my own observing nights at Keck, I received two back-to-back emails from two rival groups of researchers, each asking me to observe the same coordinates for them in the hopes of catching the fading light from what looked like a nearby burst of gamma rays. That particular event turned out to be a false alarm, debunked so quickly that I hadn't even gotten data on the target yet, but I still wonder how the night could have played out if the discovery had been genuine. If I'd gotten the data, should I have given it to the team that contacted me first? The people I knew more closely? The team who I thought would do a better job with the science? The people I might want to curry favor with for future research or employment prospects? Before the false alarm news came in, I'd gotten as far as wondering if I could take the data, hang onto it, and demand that the two groups learn to get along before releasing it to either of them, kindergarten teacher–style. Under this fly-by-night approach to ToO observations, people had been faced with these sorts of dilemmas before, and their decisions, made for any number of good or bad reasons, inevitably shaped how the subsequent science was done.

To help balance out this sort of Wild West approach to seizing telescopes, observatories today are increasingly likely to have entire systems in place dedicated to how ToO observations are handled. Observers can apply for time on telescopes that's specifically designated as ToO. In short, an astronomer can say, "If a gravitational wave and gamma-ray burst are simultaneously detected, our team will have the right to trigger a ToO to follow it up." This method is far more efficient, and in principle, it lets teams compete ahead of time for the privilege of speedy access based on the science. Some telescopes even distinguish between different degrees of ToO observations, with designations such as "nondisruptive" (a.k.a. "please point at this thing sometime in the next few days") or "disruptive" (a.k.a. "stop everything you're doing and point to it *now now now*"). Still, even in these cases, a ToO at most classically run observatories necessitates barging in on someone else's hard-earned telescope time.

The full bevy of these options was deployed in the case of the

GW 170817 kilonova. Even then, the follow-up was chaotic: people at telescopes frantically rearranged their nights, astronomers with friends who were observing bombarded every connection they could, and groups with predesignated ToO time had debates over who should get telescope access. For example, one team might be granted ToO time to "discover" the kilonova that followed a hypothetical future gamma-ray burst; after the kilonova was found, another team could argue that the first team was no longer eligible for their ToO because they were now engaged in follow-up rather than discovery observations.

The simple reality is that with limited resources and one-off events that can fade as quickly as they come, competition to be the first team to capture a ToO can be fierce. Some of this is a simple matter of science: the faster the data are captured for something like a supernova, the closer we're getting to the actual moment of explosion. The initial flash of light from a supernova can contain a unique signature of the star's outer layers and surroundings as well as the extreme physics happening deep in the star's interior, and that flash is incredibly brief. If observations are captured in those critical first moments, they can turn into a gold mine of information that we can't capture any other way.

Being fast—and therefore being first—is also simply good business. The team that can lay claim to discovering a new asteroid or getting the first groundbreaking data on a new stellar explosion may fare better down the road when requesting future funding, both because of sheer name recognition and because they have demonstrated proof that their research methods get results. Teams also don't want to pour effort and resources into being the *second* group to discover something: there's no point in publishing two scientific papers announcing identical results, and the usual rule with academic journals is first come, first served. The threat of having your exciting research scooped out from under you by another team that's faster to publish is so common in the world of academic research that it's known as just that: scooping.

In 2012, I led a team of astronomers in a mad scramble to chase down a strange star that seemed, at first glance, to be faking its own death. It originally appeared in the sky in 2009, behaving exactly like a garden-variety supernova: brightening dramatically in a few days and then slowly fading away over the following months. Observers spotted it, dutifully cataloged it as the two hundred and fiftieth supernova spotted that year—SN 2009ip—and moved on.

A year later, SN 2009ip was back, piping up with a second Monty Pythonesque flash to inform us that it apparently wasn't dead yet. The star did this twice more before a truly epic performance in 2012—when it grew twenty times brighter in only six hours—convinced us that we might finally be seeing the real deal. After the 2012 maybe-supernova, multiple teams sprang into action, all desperate to get the first definitive answer as to whether or not this star had truly, finally died. Our team quickly seized the 3.5-meter telescope at Apache Point Observatory and began taking spectra, reasoning that we could learn more about the star from its evolving chemistry than from its brightness alone. Mere weeks later, we were still crunching the data when new papers from other groups began appearing online, publishing their own hastily observed spectra.

My heart sank when I first saw the other papers; we had lost the race. It may have seemed petty, but there was an undeniable excitement to being first. It's true that scientists dream about the pursuit of knowledge, the solving of a puzzle, the moment of discovery, but it's also true that nobody's fantasy features discovering something second.

That said, the race wasn't over yet; in fact, it had turned into a marathon rather than a sprint. Nobody quite knew what to make of their quickly collected data from SN 2009ip. Everyone agreed that since 2009, we had been seeing the star during its final death throes, hurling off huge amounts of mass that, from a distance, bore a strong resemblance to the blast of material flying out of a supernova. What nobody could quite suss out was whether this 2012 event had been another stellar eruption or the real thing. Opinions were split. I thought the star must have truly exploded in 2012. Others argued persuasively that this had been just another fake-out. We kept puttering with the data and eventually published what we had in the hopes of finding any available clues as to what had happened to SN 2009ip.

Ironically, the only way to truly answer the question at this point was to wait. For *years*. As SN 2009ip faded away, my colleagues and I kept one eye on its corner of the sky, waiting to see if it would come back. When it didn't, some observers kept watching. And waiting. And watching. A decade after the odd star first burst—literally—onto the scene, nobody has seen any further sign of it, but we also still can't say for sure that it's truly gone.

The speed of scientific discovery certainly isn't unique to astronomy or even the supernova field within astronomy, and some degree of competitive effort is beneficial, causing teams to continue perfecting their methods. Still, it can sometimes stop being helpful or even cross into farcical. A few decades ago, another cosmic race was underway in radio astronomy to discover the first traces of various molecules lurking in interstellar clouds. The idea of being the first group to find, say, water or ethyl alcohol or even sugar molecules was an exciting one. ("Astronomers Find Interstellar Alcohol!" is a fun headline.) Some research groups at radio telescopes learned that teams arriving after them were peeking back through their night logs, identifying the objects or radio wavelengths that seemed to have been successful, and nabbing the same data themselves before rushing to publish and scooping the team that had made the discovery first but had been unaware they were in a race. This led to the profoundly ridiculous subterfuge campaign of people putting intentionally incorrect coordinates into their logbooks or scribbling down false wavelengths on scraps of paper and "accidentally" leaving them in the trash for the next team to find.

Simple human nature also comes into play and is, to be fair, related to why I had fantasized about my Westerbork discovery being aliens, why GW 170817 and SN 2009ip turned into a field-wide race, and why astronomers sometimes jump the gun on identifying planets as supernovae. It's exciting to be first, to be the one who spotted something, to capture that rare but incredible "Eureka!" moment in science. Even though discovery is only the first step (the science must be done carefully and correctly, not just quickly), there's an undeniable excitement to getting caught up in the rush and drama of cutting-edge science.

At the same time, the goalposts are constantly moving. Recall that the first gravitational wave discoveries were shrouded in secrecy, while today, LIGO announces gravitational wave signals on Twitter. Supernovae are also now a dime a dozen: we've discovered tens of thousands of them, so with the exception of another naked eye event, spotting a single supernova is exciting but no longer a stop-the-presses event, and the all-out sprints to study each individual stellar death have dwindled. Collecting large samples of supernovae—and keeping an eye out for the unusual ones—remains a scientifically compelling endeavor, but catching them one by one via lucky breaks and a disruptive and chaotic ToO process has become both less necessary and less efficient.

o o o

The ideal tool for this is something automated, an observing machine that can do what Oscar did: memorize a patch of the night sky and identify anything that's changed. We could then simply swing a telescope to wherever something has been discovered and get whatever data we need. Ideally, we'd also like to take those new and sudden observations without leaping on a plane or betting on another observer feeling generous that evening. Planning a lineup of different observations ahead of time or being able to observe without being anywhere near the telescope would both be massively helpful.

This method of observing is far easier said than done, but it still raises an unavoidable point. If we want to make observing more effective and efficient, chasing down events like this while still continuing the valuable day-to-day astronomical observations that don't require a rapid response, we need to start considering how to apply today's—and tomorrow's—technology to help us solve the problem.

CHAPTER TWELVE

THE SUPERNOVA IN YOUR INBOX

A gentle *click* noise from the computer let me know that my latest exposure was done. It was a little synthetic shutter sound, a handy audio imitation of the fact that the camera on the back of the Apache Point 3.5-meter telescope had just finished gathering light from my current target, a galaxy twenty-five million light-years away. This particular galaxy had hosted a mind-blowing ten supernovae in the past hundred years, bizarre compared to typical galaxies that only host about one per century. Observing the galaxy, I hoped, would serve as a forensic analysis of sorts, examining the remaining gas and dust and stars to search for clues as to why so many stars were dying there so frequently.

Exposure done, I told the telescope operator that I was ready to move, then entered the commands to nudge the telescope to my next target; we were only moving a tiny bit, to the other side of the same galaxy and another supernova explosion site. I tweaked the spectrograph setup a little bit, glanced at the guide camera on the telescope to confirm I was in the right place, then clicked Expose to start the next observations.

Once everything was up and running, I tipped quietly back in my dining room chair—I didn't want to wake anyone up—and took a sip of the coffee

I'd grabbed from the Starbucks on Eighth Avenue. It had been a quick dash down the block; I'd wanted to catch the coffee shop before they closed, it was starting to snow pretty hard outside, and while a half-hour exposure was enough to run a quick two-block errand, I still wanted to be back in time to double-check the math I'd done on how to properly move to the next supernova explosion site.

I was observing, yes, but while the telescope was in central New Mexico, I was in New York City, visiting some cousins in the lead-up to the Christmas holidays. I could control the spectrograph and fine-tune the tele- scope's position from my own laptop plunked atop my cousins' table, keep- ing in touch with the sole telescope operator in New Mexico via a small chat window. We'd been exchanging basic comments about my plans to move between targets, how the sky in New Mexico was looking (good, appar- ently, certainly better than the snowstorm starting to bury Manhattan), and whether the winds that had been blowing gypsum sand at the telescope were going to stay at bay for long enough to finish my program tonight.

I woke up later that morning after a few too-brief hours of sleep. While I had just spent the night on New Mexico telescope time, the rest of Manhattan was busy coming to life. In a long-perfected reflex, I groped around for my phone before I'd even really opened my eyes to see what emails had arrived in the middle of the night. As luck would have it, I did have a new email, from the Gemini South 8.1-meter telescope in Chile. Apparently, the previ- ous evening atop Cerro Pachón in Chile had been beautifully dark and clear, and they'd been able to get one of the red supergiants that I'd been granted permission to observe with their telescope's spectrograph. I lifted my head off the pillow and, still working on cranking open my left eye, scrolled through the email. It was great news: they'd gotten all the observations I'd requested while I'd been either asleep or operating the Apache Point tele- scope, and the data would be available to download from one of Gemini's servers at my leisure.

I'd never simultaneously observed from two telescopes before; it was slightly surreal. I had data in my hands right now from a lovely night in New Mexico and a pristine two hours on a remote Chilean mountaintop, and I hadn't even left my cousins' apartment. True, I'd be a zombie today; since I wasn't at an observatory, I couldn't easily stay on an astronomer's schedule despite working all night, and Dave and I were planning to put in a full day of work at a nearby coffee shop, run some errands, and meet up with friends before catching an evening train. I also couldn't be entirely sure until I grabbed the Gemini data whether the observations had really been done precisely as I'd described in my proposal, but still. Not bad for a night's work.

I rolled over and looked out the window, and the sight was momentarily startling. Instead of the wintry pine trees of the Sacramento Mountains or an expanse of baking Chilean summer desert, I was looking at the wall of the building across the street and some slushy snow on the windowsill.

My data from both telescopes would turn out to be pristine. I bet the skies had been beautiful that night.

o o o

The idea that an astronomer doesn't need to be physically present to observe at a telescope isn't a new one and has long been used as a way to make life easier and observations more efficient.

The earliest professional examples of remote observing stretch back to Kitt Peak Observatory in 1968: astronomers used a computer in Tucson to remotely operate a telescope on Kitt Peak, forty miles away, for several nights. In this standard model of remote observing, the astronomer was awake and actively involved with the observations as they were being taken, but they were communicating with the telescope from afar. Most remote observations happened from dedicated control rooms that had been built specifically to communicate with the telescope, with a bank of computer screens and a videoconferencing system set up so that astronomers could chat with the operators, who by and large were still sitting at the telescope itself.

Remote observing has since become increasingly common. Sometimes observing from sea level is a handy way of avoiding the physical strain of heading up to altitude. Astronomers using the Keck telescopes at Mauna Kea will observe from the Keck headquarters in Waimea, a tiny town nestled in the rolling green hills of Hawaii's northern Big Island. The perks are undeniable—plentiful oxygen, restaurants and coffee shops across the street—but the cognitive dissonance is also sometimes funny. More than one observer has had a moment of reflexive panic when hearing the rain pounding the windows in Waimea and realizing that the telescope is open; years of conditioning from classical observing can trigger a moment of "oh god, it's raining on the mirror" horror before the geography of the situation kicks in. Being in a town at sea level also means dealing with some of the drawbacks of a normal daytime schedule. The Keck headquarters keep visiting observers in their own dorm with light-blocking shades and quiet hours and generally do what they can to minimize daytime noise, but the Big Island—like most of the Hawaiian Islands—has a fairly substantial feral chicken infestation. One particularly hardy and determined old rooster has, for years, claimed the Keck visiting observer dorms as his own personal domain, and I'm convinced dozens of astronomers have been awake at 9:00 a.m. groggily googling coq au vin recipes while the rooster loudly announces sunrise to all and sundry.

Instead of overseeing the observations from twenty miles away in Waimea, observers at the University of Hawaii and some California universities, which have a dedicated share of Keck time, can do the same work from hundreds or thousands of miles away, observing from another island or even across the Pacific. The astronomy departments at these universities have dedicated remote observing rooms; astronomers who work there don't even need to leave their department in the evenings to go on an observing run.

The Apache Point Observatory 3.5-meter telescope has taken it one step further, developing remote observing software that can be installed on any laptop to allow astronomers to observe from wherever they want: their office, their living room couch, a cousin's kitchen table, and so on. Observers can work from anywhere with an internet connection. I've observed from apartments in New York, Colorado, and Seattle and even from an office in Geneva, Switzerland, while visiting colleagues for a research trip. The latter worked out particularly well thanks to the time difference. I'd been scheduled on the telescope from midnight to 5:00 a.m. in New Mexico, which was 8:00 a.m. to 1:00 p.m. in Switzerland's time zone, and it was undeniably pleasant, if also a bit odd, to wake up after a full night's sleep, make some tea, and sit down in my office to start a normal workday by opening a telescope halfway around the world.

Remote observing can certainly be luxurious, cutting out the strain (and expense and environmental impact) of traveling to the telescope. On the flip side, there are certain details lost when astronomers aren't on the mountain. Thousands of miles from the summit and the telescope, observing starts to feel a little bit like a video game, with an odd disconnect between the images on the screen and the reality of the night sky. Removed from the telescope, there's also no longer any hunting for sucker holes between the clouds; if the operator tells you it's cloudy, it's cloudy, and you can't wander outside to hopefully rearrange your night plans based on clear patches. The weather situation is sometimes even less clear. Apache Point has a weather website where astronomers can check cloud cover conditions, but this isn't always the whole story.

I was in Colorado for one remote observing run at Apache Point when I was scheduled to observe from 12:30 a.m. to 5:30 a.m. For reasons that surely made sense at the time, I decided that it was easiest for me to just stay awake all night as opposed to napping during the first half of the evening. By 11:30 p.m., I was exhausted but determined to gear up for my observing run, and according to the weather website, the sky was, somewhat surprisingly, clear. Ready to snap myself into gear, I made myself a large pot of strong espresso and downed the whole thing, then sat down at my laptop and logged into the remote observing software with a resting heart rate in the 130s and a serious excess of caffeine-propelled energy.

The telescope operator jumped onto the chat almost immediately. "Hi, Emily. So it turns out the summit is completely socked in with low fog. I'm betting we won't even be able to open tonight. Why don't you give me a good number to reach you at and go to bed? Get some sleep, and I'll call you if things improve." Since I'd just slammed enough caffeine to jump-start a wooly mammoth, I was left twitching in my living room, geared up for an observing run that was almost certainly not going to happen. Lesson learned: coffee *after* confirming the night is on.

Sea-level emergencies can also become a problem. Observing from your living room is all well and good…until your internet goes down. I remember a panicked midnight bike ride from my apartment one evening during another remote observing night at Apache Point after my internet unceremoniously dropped out in the middle of an observation. To the operator, it looked like the astronomer had essentially vanished mid-exposure, and I wound up pedaling frantically to my office, which I knew would have reliable internet at 2:00 a.m., to try and lose as little observing time as possible. Other observers have been stuck in snowstorms or evacuated from their buildings thanks to fire alarms while the distant telescopes they're observing with sit happily but motionless in perfect weather, waiting for the remote astronomer to get back in touch.

Another disadvantage of remote observing is that astronomers don't get to fully duck out of everyday life. It may be an ordeal to get yourself to an observatory summit, but once you're there, there's a certain delightful isolation: you're there to observe and that's it. By contrast, most professional astronomers find it practically impossible to fully detach from day-to-day life for observing runs when they're still embedded in it. When someone is observing from their office up the street or their laptop on the kitchen table, they're often as not closing the telescope and then heading home to help get their kids off to school or gearing up for the normal workday responsibilities that are still on their plate during daylight hours. I've taught classes while glassy-eyed from a full night of observing and two hours of sleep, and one colleague described reading her kids bedtime stories during long exposures at the start of an evening.

Amid the day-to-day normalcy of your own couch or office, it's all too easy to lose track of the reality that your occasional clicks and keystrokes on a computer screen are actually physically *moving* a many-ton instrument thousands of miles away. This is part of why Apache Point Observatory insists remote astronomers get trained in person before they can observe using the remote software. The convenience and accessibility of remote observing are undeniable, but gathering data from afar without ever going to telescopes can lead to a slow but persistent disconnect between astronomers and the tools they use. Opportunities to learn the ropes of a world-class telescope in person are becoming increasingly hard to come by.

o o o

Remote observing allows astronomers to use telescopes from afar, but the astronomers are still an active part of the process in real time. In this model, the observer who will ultimately be working with the data is the one starting and stopping exposures, working through or shuffling a target list as needed, and staying awake and engaged for as long as the telescope is in their however-distant hands.

Another approach to remote observing is a technique known as queue observing.

Astronomers already arrive at the telescope well-prepared: proposing for telescope time requires that good observers have thought through which targets they want to observe and when, which telescope instrument they want to use and how it should be set up, how long they need to expose, and even the rough order of observations they'd like to carry out. It's true that a lot can be gained by taking an image, examining it, and then trying again on the fly to improve the data quality, but oftentimes, observers' preparation is so thorough that the data comes in exactly as expected on the first try. When all is going well, an observer might not be doing much more than going through a checklist and clicking buttons, opening and closing a shutter, and moving from object to object in precisely the order they'd intended. It's not too hard to imagine taking the next logical step: if the astronomer doesn't have to be physically present, and if everything has already been carefully planned ahead of time, does the astronomer really need to be involved in the actual observing at all?

In queue observations, astronomers prepare every last detail of their observing program months ahead of time—what objects they want to point at, for how long, and exactly how they want to configure each element of the telescope—and assemble it into a step-by-step plan. An observatory can then combine the plans from all astronomers who have been granted time and use them to assemble a queue, or ordered list of planned observations, that's designed to optimize the precious hours of available telescope time.

The method opens up a wonderful range of possibilities. Rather than observing one astronomer's entire program at once, during a night or stretch of time when they've committed to being awake, and then jumping to the next person on the schedule, a queue can mix and match observing plans, grouping requests from different astronomers based on where their targets are in the sky or which instruments they want to use. Observations from one astronomer's program that require several hours of long exposures can be paired with short exposures from another program to squeeze every second of science possible out of a typical night.

The Hubble Space Telescope takes this approach for obvious logistical reasons. (Sadly, observing with Hubble does *not* mean suiting up and heading to a launchpad for a personal trip to the telescope.) Astronomers who get time on Hubble—which is distributed in orbits rather than nights—are asked to prepare intricate observing files laying out every last detail of how the telescope should be configured and how every single second of each

orbit is going to be spent. Simulating and plotting out the observations can take weeks, especially for new observers, and the plans must be submitted by a hard deadline in order to make it into Hubble's queue. The first time I was awarded Hubble time, I was ecstatic…and also getting ready to leave the country and my work laptop behind for a solid month. After eleven years of dating, Dave and I had gotten married just a few days earlier, and the email approving my proposal arrived hours before we left for our honeymoon, a trip we'd been planning for years that stretched right through the deadline for the observing plan. Still, faced with the news that I'd get to use Hubble for the first time, Dave immediately got the observing software downloaded and running on the bare-bones shared laptop we were bringing with us, I gathered all the Hubble user manuals I could find (yes, even telescopes have user manuals—*many, many* manuals), and I spent one day of our honeymoon frantically finishing up my observing plan from a hotel room in Istanbul.

This style of observing may require even more advance planning than a typical observing run, but it's so efficient that some ground-based telescopes have adopted queue observations as well. For our ground-based observatories—like the twin Gemini 8.1-meter telescopes in Hawaii and Chile—this method also minimizes the telescope time that's lost to weather, using a standing list of queue observations to pair astronomers' proposals with their optimal weather conditions. A staple of classical observing has long been that if you go to the telescope and your assigned night is cloudy, you're simply out of luck; your assigned time comes and goes, the telescope sits idle, and there's no option to try again on a better night. Queue observing is more flexible. If a night is slightly cloudy, a queue-based telescope can tee up an observing plan from someone with observations that can tolerate a less-than-ideal sky, and the person who needs clear weather is put back in line to wait for a better night.

It's an undeniably luxurious way to observe. Waking up to a bundle of new data in your email inbox is certainly more restful than powering through

the 3:00 a.m. haze and patchy clouds in the hopes of maybe squeezing out whatever scraps of data you can manage on your preassigned night. At the same time, queue observing removes the astronomer from their observations by one more degree.

To be fair, this isn't always a bad thing; consider the many sleep-deprived mistakes by tired observers chronicled here. In astronomy, just like anywhere else, there will always be skeptical Luddites—even back in the days of photographic plates, some observers would scoff at having night assistants or students take their data for them, declaring that they simply couldn't trust data they hadn't taken themselves—but in reality, telescope operators already know the telescopes and instruments better than many astronomers, and putting well-laid plans into the hands of an expert can often go extremely well.

On the other hand, this means that the person doing the observing is one step further removed from the science. A mistake in the queue, the instrument setup, or where the telescope is pointed could just as easily have happened to the astronomer themselves as to the telescope operator and trained observing staff, but it's still hard to swallow for an astronomer getting suboptimal data back from queue observations. There's something to be said, however stressful it may be, for the person doing the observing having a solid personal investment in how it turns out. Even with queue telescopes operating like a well-oiled machine, I've gotten back Gemini observations that were pointed ever-so-slightly to the wrong part of the sky or that were set up just a tad differently than I'd requested. The difference is small, maybe even imperceptible, to someone following the instructions I put into my queue observing plan, but it can make a world of difference when the data actually arrives.

It's still *possible* for astronomers to visit queue-based telescopes and be present for the observations—some detail of the observations might be tricky enough to require the astronomers themselves to be present—but even then, the strict policies of queue-based observing remain in place. I've

observed classically at Gemini South in Chile and Gemini North in Hawaii, but in both cases, I still had to submit a precise list of targets and exposure times and weather requirements ahead of time, and the list was extraordinarily difficult to change. A few days before traveling to Chile to observe a handful of galaxies at Gemini South in person, I'd come across a newly discovered galaxy in a recent research paper that would have been a great addition to my list. I logged into Gemini's software to add it to the queue and was promptly informed that such a change was strictly not allowed: my program had been approved based on my original list, and I wasn't permitted to add targets. It's the sort of change that could have been easily made if I'd been at another telescope, but on Gemini's queue system, all the targets had to be preapproved.

Once I got to the telescope, I was largely relegated to the role of spectator, watching as the telescope operator and scientific staff executed the steps of my program for me. I was there, able to step in and ask for a small tweak to the instrument setup or a nudge to make sure the telescope truly was positioned properly on the sky to catch my galaxies, but for the most part, I served as a sort of supervisor to make sure the observations went according to plan. I spent a decent chunk of the time feeling a bit useless, and other observers at these telescopes described a similar feeling of being at loose ends. I'd heard from someone that another queue-based telescope in Chile prevented visiting astronomers from so much as touching the telescope controls: everything was done by the operators and telescope staff. Supposedly, they'd installed a switch labeled "astronomer" for visiting classical observers to use that could be flipped back and forth but wasn't actually attached to anything. Its main purpose was to give the people who'd insisted on being present for their observations something to do.

Luckily, the weather on my classical night also agreed with the preprogrammed weather constraints I'd attached to my program. If it hadn't, we would have moved on to a different program in the queue. This could have been great; if I'd had thick clouds or very bad seeing, the telescope would have been busy taking data for someone else who could use it rather than just sitting there and waiting in case the sky improved. On the other hand, it could have been disappointing; I didn't need picture-perfect sky conditions, and a colleague of mine had once traveled to Hawaii to observe during a beautiful night on Gemini North and then promptly lost his time. The sky had been so excellent that another program in need of particularly pristine conditions got bumped ahead of him in the queue, leaving him to twiddle his thumbs while the telescope observed completely different targets.

When we finished my classical Gemini observations about an hour early—I'd had to cut one galaxy from my list because it was in a cloudy patch of sky—the operators again turned to the queue and began observing a completely different program. Logically, I understood that this was good—I'd already gotten my data and didn't strictly need that last hour of time, and somewhere, another astronomer would wake up the following morning and be delighted to see new data for their program in their inbox—but it was still strange to be summarily shut out of my own observing night by a computer program. The queue approach at the telescope had gained a satisfying degree of efficiency—it could constantly chug away on a curated list of carefully preplanned proposals—but it had lost a bit of the nimbleness and creativity that astronomers had come to enjoy when a telescope was placed in their hands for a night of their own.

o o o

Every astronomer is curious about their own corners of the night sky, designing observations aimed at answering specific questions. Still, the fact remains that plenty of astronomical observations can, in their execution, be surprisingly similar. For imaging in particular, there are several standard sets of filters that astronomers use—letting only blue or red or infrared light through to the camera—and fairly simple requirements for capturing those images, exposing for long enough to get a strong signal but not so long that bright objects saturate the camera's detector. Someone asking for observations to observe a young, bright, distant star cluster in one part of the sky is probably describing a sequence of telescope commands that are practically identical to someone asking for observations of an old, dim, nearby star cluster in *another* part of the sky.

For observations like this, we don't need astronomers at all, at least not for the initial data-gathering phase. Astronomers may *use* the data once it's available—to measure the position or brightness of a star, map a star cluster or galaxy, or look for the tiny flash of a supernova or the smudge of a passing asteroid in a well-mapped area of the sky—but really, nobody in particular needed to request or specify the observations. If everyone wants the same sorts of images of different patches of sky, we could potentially ask a telescope to simply go get them, right?

Robotic telescopes have started to fulfill this purpose in the past couple of decades, often with great success. A robotic telescope can be sent to a particular patch of sky to take a standard set of observations, a process that cuts out the concept of an observer almost entirely. Some of these telescopes take requests, like the Las Cumbres Observatory Global Telescope network. Combining twenty-five 0.4-meter, 1-meter, and 2-meter telescopes scattered across the world with an artificial intelligence scheduler, the network gathers requests for observations from astronomers along with other data, such as weather at the different telescope sites, and then directs its telescopes to capture the observations before compiling the resulting data and sending it back to the interested scientists. The Las Cumbres network can repeatedly image a single star or patch of sky, chase new discoveries as they happen, and even automatically take spectra of some targets.

Other robotic telescopes have their science preprogrammed, following a sequence of predetermined moves throughout the night to survey the sky. Some may look at a patch of sky just once, capturing an image that can then be studied by curious astronomers (or used to spark more tailored follow-up observations), while others may look at the same piece of sky over and over,

looking for anything that might have moved or changed. The latter are especially adept at capturing new supernovae or moving asteroids or even stars
that vary more subtly. By measuring their brightness repeatedly for months
or years, we can identify patterns and periods in the stars' variations, and
observations like this are much more efficiently done by a telescope robot
than by an observer who might need endless nights of telescope time to
capture the same images over and over by hand.

The observations may seem too dull and simple for human beings to
bother with, but the science that can be done with them is anything *but* dull.
Mike Brown described a robotic survey searching for Kuiper Belt objects—
similar to the work Dave Jewitt and Jane Luu had done by hand early on,
when the Kuiper Belt was first being studied—that used the recently robot-
icized forty-eight-inch telescope at Palomar Observatory. The telescope
would observe all night, taking three images of each patch of sky it studied
to look for anything that moved, and then transfer the data to Pasadena,
where Mike worked. A computer could even analyze most of the data rather
than forcing a person to flick through it by hand. In the era of digital data,
astronomers are increasingly able to program automatic reduction and analysis software—"pipelines"—to analyze their data. After all, if the setup and
work of the telescope is standard, the methods for analyzing the data will
be fairly standard too. The program used for the Kuiper Belt object search
would throw out anything stationary and keep anything that looked like it
might be moving, which Mike could then review by hand. It was purposely
programmed to be generous, including false positives rather than risking
throwing out data that could be real, so on any given morning, there were
usually one to two hundred potential moving objects for Mike to check.
Every couple of days, a bona fide Kuiper Belt object would crop up, and
each one would come with a small frisson of excitement.

On one January morning in 2005, Mike was sorting through the
moving object candidates and came across a shockingly bright and slow-
moving target. After the initial skepticism familiar to any scientist ("what

did I screw up this time?"), Mike began digging into the data and scribbling notes, slowly realizing as he worked that he was looking at a real and absolutely huge distant object in the Kuiper Belt. The discovery turned out to be Eris, the largest Kuiper Belt object ever found at the time. Though its size was ultimately downgraded (it's smaller than Pluto by about fifty kilometers), Eris's discovery prompted an infamous vote by the International Astronomical Union to better define what astronomers considered a planet. The vote demoted Pluto to dwarf planet status, along with Eris and several others, a story detailed by Mike in his book, *How I Killed Pluto and Why It Had It Coming.*

Robotic telescopes, of course, aren't always perfect, and teaching a telescope to observe by itself comes with its own host of problems. Mansi Kasliwal described efforts to roboticize the forty-eight-inch telescope at Palomar, which included an automatic arm that could swap out filters in front of the telescope's camera. Operating directly above the telescope, a malfunction in the arm risked dropping a filter onto the mirror itself; the arm did once fail, but the filter was fortunately caught by a second fail-safe mechanism installed with precisely this sort of error in mind. The same automatic filter arm also had to stay warm enough to operate effectively in the midst of a telescope that was otherwise kept quite chilly, leading someone to design what was essentially a cozy robot mitten for the arm (apparently even robots get cold while observing in the dome all night).

The Palomar forty-eight-inch also had an old safety measure that involved sounding a loud horn in the dome before the large and heavy telescope turned itself, then waiting for thirty seconds (presumably so anyone within earshot could run away) before swinging itself around. While it's an understandable safety precaution to take, Mansi pointed out that at a robotic telescope that requires no human intervention, they could simply lock the door to the dome while the telescope was observing to prevent problems. Other robotic telescopes have, in the past, swung themselves into protruding parts of the dome or accidentally pointed themselves directly at the floor while they've been learning the ropes of how to observe alone.

On the other hand, the perks of robotic telescopes are undeniable. When properly programmed and operating, they can observe immense swaths of the sky entirely without human intervention, leaving the humans to explore the science contained within the data rather than laboring through the repetitive and tiring efforts required to get the data in the first place.

Robotic facilities haven't entirely left the art of in-person observing behind. The people who design and program these telescopes certainly still need to be conversant in the details of how observing works. It's also true that while the role of astronomers at the telescope may be diminishing for some types of observing, there are other types of observing that still require humans to be active and present. Still, even in astronomy, there are always people who will grump about computers and automation, complaining that removing the astronomer from astronomy and letting the machines take over is going to be detrimental to the science.

Proponents of robotic telescopes and automation in observing argue that these innovations free up the trained astrophysicists to do the jobs that robots can't. Today's astronomers have the option of carrying out entire research projects—entire careers, even—that make use of telescope data without ever visiting or running the telescope themselves. The sheer volume of data taken by telescopes like this is also growing immensely. As remote, queue, and robotic telescopes continue to grow in popularity and capacity, the very nature of astronomy and observing is beginning to change.

CHAPTER THIRTEEN

SYNOPTIC FUTURE

I wake up when the shuttle bus turns off the pavement and onto the dusty dirt road that leads up to Cerro Pachón in Chile. Like so many Chile trips I've taken before, I tip my head against the rattling window and take in the stark scene outside. The base of the mountain is shrouded in thick mist, the dusty ground and scrubby plants emerging in a rolling sameness along the side of the road as we drive. The fog is unusual, but that could be due to the time of day. In a notable diversion from my other telescope visits, I'm making the trip up the mountain at 6:00 a.m., arriving in time to eat breakfast—normal morning eggs and toast and coffee, with a crowd of construction crew members—at the cafeteria. I'm not night-shifted right now because I don't need to be; I'll be heading back to La Serena on the afternoon shuttle. I may be visiting an observatory, but I won't be staying on the mountain tonight.

When we finally reach the observatory itself, we've popped above the fog, and the shuttle drives right up to the door of the service and operations building. The structure towering above me as I step out into the sun is gargantuan, a long, futuristic-looking building topped at one end by the steel-strutted skeleton of what will eventually be the telescope dome. During my visit, the dome is being built, but equipped with a vest and hardhat and steel-toed boots—the same ones I wore fourteen years earlier when I was a tour guide at the VLA—I'm taken up to walk around the dome itself and even stand in the center where the telescope itself will be. Teal-painted metal all around me traces the incomplete outline of a huge blocky dome and shutter system, currently tangled up with construction scaffolding and framing an empty round stage of concrete. As I stand here, the 8.4-meter mirror that will eventually be the star of the dome is en route to Chile, a two-month trip that includes several specialized heavy-transport vehicles and a trip through the Panama Canal. For now, I can look straight through what will be the walls of the dome. Turning in a circle, I can see the Gemini South telescope further down the ridge of Cerro Pachón and, in the distance, the cluster of telescopes at the observatory atop Cerro Tololo. It strikes me that with one spin, I'm seeing the full spread of how we currently observe in astronomy: classically operated telescopes, robotic telescopes, Gemini and its queue-based system, and the future being built in this very spot.

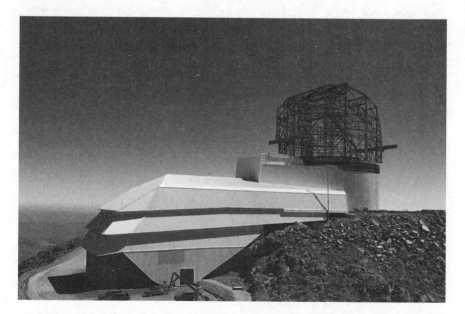

The under-construction Vera Rubin Observatory in March 2019. *Credit: Emily Levesque.*

This telescope will soon become one of the most powerful observatories of the 2020s. Nicknamed LSST for years—short for the Large Synoptic Survey Telescope—the facility began the new decade with a new name honoring one of the great observers of decades past: I'm standing at the site of the Vera Rubin Observatory.

Downstairs, the service and operations building is sleek and bright white in the sun, with several stories of windows and subtle angles that make it look a bit like an odd desert yacht. It may look stylish, but the design is a purely practical one; the sloping angles of the building have been designed with the wind currents on the summit in mind to provide the least disturbance in airflow for the telescope. The building itself houses everything the telescope will ever need: a room for telescope operations, two clean rooms for working on the telescope's state-of-the-art camera, and an enormous elevator and rail structure so the mirrors can periodically be removed from the telescope and transported several floors down and over to a custom-designed coating chamber. The chamber will recoat the telescope's mirrors every few years with aluminum and silicon-protected silver. Scattered around the immense work area are several giant drums of ballast (recently used for testing the elevator that will carry the mirror), a blank for the telescope's secondary that was used to test a mechanical support system, and, locked carefully away in a sealed metal box, the actual secondary mirror for the telescope, recently delivered from Rochester, New York, and waiting to one day be installed.

The service and operations building also includes a smattering of offices and a small conference room where the project leaders back in La Serena or the United States call in to catch up with the leaders of the various teams currently working on the mountain during a daily morning meeting. During today's meeting, I'm briefly introduced—"This is Emily. She's an astronomer writing a book about observing."—and then quickly lose the thread since the rest of the meeting is conducted in Spanish, efficiently running through a massive spreadsheet of the day's teams and tasks. Instead, I take in the place where we're all sitting. The building is shiny and new: simple but tasteful light wood floors, white cabinets, more teal accents, and a distinct *eau de IKEA*, right down to gleaming long, low windows with shades that could be raised via remote controls and levered openings…but those windows looked out onto the astonishingly barren dust-red expanse of the Andean foothills. It's a fascinating mismatch; windows like this should look into a neighboring building on a college campus or a little patch of lawn in a suburban office park, not the edge of the world.

While the dome continues to march toward completion, work is also underway on the building's support facilities. One group is running coolant lines to the clean rooms, another is doing tests on the just-finished coating chamber, and a third group is working on what looks like a minor hiccup in the building's backup generator systems. To fix it, they explain, they'll need to shut down power to the whole mountain at around noon. The phrase *shut down power to the mountain* is a handy reminder that despite the office chairs and teleconference equipment, we're in the middle of nowhere. It also explains why teams are building clean rooms and coating chambers on-site: if the telescope's camera needs repairs or the mirror needs a touch-up, the work will need to happen here. As complicated as it is to build a self-sustaining facility, it's still cheaper and more efficient than trying to transport the camera—or, even worse, the mirror—to another site for maintenance.

The true gem of the mountain is its network and data capability. An entire room in the building is dedicated to rows upon rows of server racks, strapped to each other and the floor to prevent earthquake damage and equipped with their own independent cooling and fire-prevention systems. When finished, the telescope on Cerro Pachón will be connected to the base facility in La Serena by a fiber network that can transmit six hundred gigabits of data per second. At that speed, the telescope could transmit the entire *Lord of the Rings* film trilogy—extended edition—to La Serena in less than half a second.

It's an immensely impressive facility, but this effort goes beyond just building another excellent telescope. There's a reason the Rubin Observatory will have the most advanced camera in the world and a network that will rival that of most tech companies.

The Rubin Observatory's science mission is simple, straightforward, and spectacular: it will image an enormous swath of the southern sky, once every few days, over and over and over, for ten years. With its 8.4-meter mirror, the Rubin Observatory will be capable of detecting objects as faint as anything ever observed from the ground and will ultimately generate the equivalent of a decade-long movie of the entire southern night sky, observing billions of objects for the first time and tracking how they change on an unprecedented scale.

All this observing will also generate an embarrassment of riches when it comes to the data. Thanks to the Rubin Observatory's camera, one single image from the telescope will be 3.2 gigapixels in size. Displaying that single image at its full resolution would require fifteen hundred high-definition televisions. A single night of observations with the Rubin Observatory will generate thirty terabytes of data, and the data management tools for the facility will include nearly instant alerts and updates as the telescope spots anything in the sky that has moved or changed.

The sheer volume of science that the Rubin Observatory will produce is almost impossible to contemplate. Imaging the sky over and over, it's expected to detect over one thousand new supernovae in a single night (right now, we detect fewer than one thousand supernovae in a *year*). The Rubin Observatory will find asteroids and other moving objects in our solar system, including those classified as near-Earth objects that could potentially come close—even hazardously close—to our own planet. The telescope will track every star in the sky whose brightness varies over time, giving us a full decade of constant data on stars as they evolve, change, and, in some cases, approach the ends of their lives.

The Rubin Observatory will also manage all this with almost nobody on the mountain at all. Operators will be present, true, but they'll mainly be serving as a steward or supervisor of the telescope, getting things up and running in the evenings and shutting down in the mornings and keeping an eye out in case anything goes wrong. Aside from operators and a few other staff members, the mountain will operate on a skeleton crew. Team members will make occasional daytime visits to make sure everything is working as it should, but for the most part, the Rubin Observatory will be empty.

During my visit, the site is far from empty. Various construction crews are all over the place, working in parallel to complete the many different moving parts of this observatory in anticipation of its first light—when a new telescope is opened and takes an image for the first time—in 2020. During my visit, I come to the slow realization that this immense amount of work is all for what will ultimately be an almost empty facility, ready for problems and equipped for visitors and carrying out incredible new science but with only a sparse handful of humans needed to keep things running. I'm watching people build a place that has been very purposefully designed as a ghost town. Once the Rubin Observatory is up and running, any little long-whiskered viscachas on this particular Chilean mountain will likely be watching the sunset alone.

o o o

The Rubin Observatory is one of the crown jewels of a new era in astronomy, where observers are able to harness the power of automation to turn telescopes from night-to-night tools used by individual astronomers into veritable science factories.

The idea isn't a new one; in fact, the phrase *science factory* is precisely how project manager Jim Crocker described the Sloan Digital Sky Survey telescope, a predecessor of the Rubin Observatory that began surveying the northern sky in 2000. Built at Apache Point Observatory in New Mexico, it had all the same nemeses as the other telescopes on the mountain— the blowing gypsum sand, the ladybug swarms, the scourge of the miller moths—along with the challenge of running a massive survey operation. Its 120-megapixel camera produces about two hundred gigabytes of data every night, and over the years, it has taken deep multicolor images of a third of the entire sky and observed spectra of more than three million objects. The book *A Grand and Bold Thing* by Ann Finkbeiner details the years of effort and collaboration and innovation necessary to build a survey telescope of this magnitude. In it, Finkbeiner notes the immense volume of data produced by Sloan's techniques and the fundamental change it has wrought in how we do astronomy. Rather than a handful of photons or stars or galaxies, astronomers using Sloan data now have millions at their disposal, and the challenges for some research areas have shifted from observational—battling to extract science from dim faraway objects—to computational, asking and answering scientific questions with the riches of an infinite universe. We're facing another shift of this kind with the Rubin Observatory.

Of course, even with all its impressive planned-for accomplishments, everyone recognizes that the Rubin Observatory can't do everything. It won't, for example, be equipped with a spectrograph during its initial ten-year run; every moment of its time will be taken up with imaging. It may be equipped with the best camera in the world, but it won't be able to detect light outside the narrow range of visible wavelengths. It won't be launched into the upper atmosphere or dragged to the site of a future Venus transit or asteroid occultation.

As a result, no matter how impressive the Rubin Observatory will be, it will also benefit from not working alone. It'll be detecting a thousand supernovae per night, true, but at least a few of those supernovae will need to be studied more closely with other telescopes, and the same holds true for anything else new and surprising that the Rubin Observatory finds. Astronomers are still sorting out the best way to manage ToO observations and proposals for observing time and queue programs in this upcoming new era, but the fact still remains that while the Rubin Observatory will be cranking out hundreds and hundreds of new discoveries, many of its discoveries will be only the first piece of the puzzle, with other telescopes and other astronomers pursuing the new science that the Rubin Observatory has revealed.

o o o

The most exciting scientific discovery of my career almost didn't happen.

In September 2011, I was back at Las Campanas Observatory—the same place where I'd lost an entire visit to wind years before—to observe what I'd identified as some very strange red supergiants. A few years earlier, Phil and I had published some research pointing out a few oddly behaving stars. They seemed to change their temperatures bizarrely fast, on a timescale of months, far too fast for stars that were ten or twenty times as massive as our sun and as large as the orbit of Jupiter. What's more, they were going from cold to too cold, reaching temperatures so cool that they flew in the face of everything we knew about the physics of stars.

Our weird stars caught the eye of an astronomer named Anna Żytkow, who emailed us with an interesting idea. A few decades earlier, she explained, she had worked with Kip Thorne—the same Kip Thorne who went on to win a Nobel Prize for the detection of gravitational waves—on a theory for a completely new type of star. This star, they predicted, would outwardly look almost indistinguishable from a very red and very bright supergiant. However, rather than the typical core that supported pretty much every star we knew about, generating energy through nuclear fusion, Kip and Anna's star—a Thorne-Żytkow object—would have a neutron star for a core. Instead of fusion, the star would essentially be supported by quantum physics despite looking like a completely normal star to a casual observer; the only sign of its bizarre interior structure would be odd but subtle chemical excesses lurking at the star's surface. If Kip and Anna were correct and Thorne-Żytkow objects did in fact exist, they would represent a completely new model for how the insides of stars could work.

Thorne-Żytkow objects had been proposed back in 1975, but for more than thirty years, nobody had ever seen any evidence that they existed. A few research groups had carried out searches but found nothing definitive. Still, with our expertise on red supergiants, we were in a great position to take up the search again, and Anna thought our odd, cold, variable stars could be excellent candidates. The idea fascinated me. Thorne-Żytkow objects sounded like an enticing chimera—the weirdest stars in the universe, neutron stars, combined with the biggest, red supergiants—and the idea that we could potentially find the first ones seemed too good to pass up.

Excited, I drew up a telescope proposal, but I knew that "Hey, we're looking for a bizarre new type of star that's never been seen before because we have a good feeling about this!" didn't make for a very scientifically robust argument. We decided this first observing run would simply be a preliminary search: we'd assemble a list of cold, bright red supergiants and take detailed spectra of each star one by one, measuring their chemistry to establish a baseline of what was "normal" in stars like this. The odd elements we expected to find in excess in Thorne-Żytkow objects were very specific and esoteric—lithium, rubidium, molybdenum—and unsurprisingly, nobody had ever sat down and made an exhaustive study of how much molybdenum was in a normal red supergiant, let alone an abnormal one. Before we could find the oddballs, I reasoned, we had to survey the ordinary stars. Still, a dull old chemical survey wasn't very exciting either, so the final proposal was a compromise: we'd be surveying about a hundred red supergiants, including a large sample of normal ones along with a few weirdos. The proposal succeeded, and we were granted three spring nights on one of the 6.5-meter telescopes at Las Campanas.

It was a fun and quirky project and a bit unlike any research we had done before, so on the first night of our observing run, we were still fine-tuning our plans. As the afternoon wore on and we geared up to begin observing, Phil suggested adding a few more stars to our list, using old imaging data to identify a few extra-red supergiants that could be excellent Thorne-Żytkow object candidates thanks to their extreme colors. We were also observing with another collaborator of ours, Nidia Morrell, a brilliant astronomer with decades of experience at Las Campanas who knew every telescope, camera, and spectrograph on the mountain backward and forward. She proposed a few last-minute tweaks to the target list order and instrument setup that would help make our observations as smooth and efficient as possible.

During our observations, the data that came in was largely gibberish. After the run, I would spend a few solid weeks poring through manuals and software documentation to work out how to properly reduce the data, peeling away all the junk signals from the sky or the electronics or other nearby stars to pin down the light from each star we were interested in. Still, there was just enough detail popping up on the screen as every exposure came in for us to get a rough look.

Nidia, of course, was the exception; like all the other instruments on Las Campanas, she knew this one inside and out and could recognize little quirks in the data immediately as it came in. Sometimes, she could point out a problem—a star that hadn't been perfectly centered, an adjustment we could make to improve the signal—and at other times, she would simply point out small details or features, helping us to recognize bits of science peeking out of the data even in its raw state. Most of the data looked like a stack of thin white horizontal stripes, with the occasional dark gap in a stripe here and there showing a break where one element or another was absorbing light at a very specific wavelength.

Partway through the night, we came to one of the new extra-red stars Phil had added to the list. I'd pulled its name out of an infrared catalog and listed it in my notes for the evening as the unpoetic J01100385–7236526, a basic amalgam of coordinates indicating precisely where it was in the sky. We took an exposure and then curiously watched the data appear on the screen as we prepared for the next target. It looked good—clear and bright—but we immediately homed in on something unusual in the white stripes. Along with the usual dark gaps along the stripes, there were several atypically bright white dots, clearly showing spots where some specific element in the star's atmosphere was emitting extra light. This was weird and something we certainly hadn't seen before in our red supergiants; elements in the atmospheres of stars like this were typically only *absorbing* light, not *emitting* it. That said, nobody had ever predicted that Thorne-Żytkow objects would emit light like this either, so it wasn't a "Eureka!" moment or a clear "Aha!" or even terribly exciting at all. It was a very subtle "…huh." Everyone's heads tipped from side to side as though to make sure the weird bright spots were still there, and we didn't say much at first.

Nidia was the one who finally piped up, possibly seeing something we didn't in the odd data that had popped up on the screen. "I don't know what it is, but I know that I like it!"

I wrote what she'd said next to the entry for J01100385–7236526 in my notes and moved on.

Over a year later, I was steadily working through the data, now using the stars' proper names, and came across a star named HV 2112. To my utter shock, HV 2112 was an aberration in our collection of red supergiants, displaying bizarrely excessive amounts of a few key elements. After running the math and comparing it to every other star in our sample, it was clear: HV 2112 had much more lithium than the other red supergiants in our sample. It also had much more rubidium. And much more molybdenum. In short, it had exactly the same strange chemical profile that had once been predicted for a Thorne-Żytkow object.

HV 2112 was also doing something else weird: it was emitting light from hydrogen atoms in its atmosphere. Nobody had ever predicted this behavior for a Thorne-Żytkow object, but a little digging turned up something new. Some stars could occasionally wind up with hydrogen emission in their atmospheres thanks to the energy generated by stellar pulsations: if the outer layers of the star were unstable, they might occasionally pulse like odd gigantic heartbeats, creating a distinctive pattern of light emitted by hydrogen in their atmospheres. Nobody had ever predicted that

Thorne-Żytkow objects would show signs of hydrogen emission, but a few people *had* predicted that their outer layers could be unstable and that they might pulse.

HV 2112 had just become the best candidate ever discovered for proving Thorne-Żytkow objects were real. It meant we could have the first evidence in hand for a completely new model of how stars could work. It opened up heaps of new questions: How did Thorne-Żytkow objects form? How long did they live? Could they make black holes when they died? Could they make supernovae? Could they make gravitational waves? Could they make something we'd never seen before? An entirely new realm of science questions had been opened up, all thanks to the data we had captured on HV 2112.

As I stared at my computer screen, the pieces of the puzzle assembled in front of me, the hydrogen emission detail rang a bell. I grabbed my observing notes and flipped back to that evening, when Nidia had spotted emission in our data from J01100385–7236526. A quick trip to a database of stellar coordinates and names and it was confirmed: J01100385–7236526 was HV 2112, the potential Thorne-Żytkow object. A star we almost hadn't observed at all, that made it onto our list at the last minute thanks to our classical night of observing time and Phil's eleventh-hour tweaking of our proposed plan. A star that had caught the eye of a veteran observer almost immediately as a quiet harbinger of things to come.

o o o

The shift that astronomical observing has experienced over the past half century has been nothing short of breathtaking. We've moved from an era in which astronomers had to fight for photons, hunching in prime focus cages and gathering data on glass plates, to an era where giant robotic telescopes can produce epic heaps of data. It's a scientifically exhilarating thought: the scale of our research is inching closer to the scale of the cosmos, and every time we've pointed new technology at the sky—larger telescopes, airborne telescopes, telescopes with their vision sharpened by lasers—we've discovered something new and unexpected about our universe. This volume of data is also a wonderfully democratizing tool in astronomy. Far from the era in which observational data was reserved for people who had access to the handful of research-grade telescopes (whether thanks to their university affiliation, their research funding, or their gender), data from these telescopes can be made available to anyone. Astronomers all over the planet will be able to share in the wealth of data produced by these immense surveys.

Seeing how much more data can be gathered by one telescope, one might be tempted to think that we simply need fewer telescopes today. If the Rubin Observatory can do in a night what an observatory like Palomar used to do in years, can't we close down Palomar? If we can build a telescope that generates thirty terabytes of data in a single night, won't this be enough data for every astronomer on the planet? Can a single powerful telescope be so spectacularly built that it meets the needs of everyone and answers every question in astronomy?

The answer is a resounding no. For one, collecting new discoveries isn't enough; we need to explore what we find, and the Rubin Observatory is exciting in part because of the other telescopes it will complement. Two enormous telescopes in the Southern Hemisphere will also achieve first light during the 2020s: the 24.5-meter Giant Magellan Telescope and the 39.3-meter European Extremely Large Telescope, both of which will join the Northern Hemisphere TMT to become the largest optical telescopes ever built. As the Rubin Observatory discovers supernovae, asteroids, and distant galaxies with its 8.4-meter mirror, larger telescopes are crucial for studying these new discoveries in detail. Since taking a spectrum of an object requires spreading that object's light out according to wavelength, taking a good spectrum of a dim object requires a much bigger mirror than taking a good image. Because of this, the two enormous telescopes planned for the Southern Hemisphere are the only ones that will be able to directly dig into the objects the Rubin Observatory discovers, measuring their chemical compositions and distances. No matter how well it's built, the fact also remains that the Rubin Observatory from its home in Chile, can't look through the planet to study the northern half of the sky. We'll still need telescopes in the Northern Hemisphere, and lots of them, to help us answer the new questions about our universe posed by the Rubin Observatory.

A massive survey telescope like the Rubin Observatory will discover millions of supernovae, but the extremely large telescopes of the coming decade will let us study the chemistry of those supernovae and explore their host galaxies to study the stars that made them. Airborne and space-based telescopes open up the wide world of light hidden from us by our planet's atmosphere, and radio telescopes reveal the potent science held by light far beyond the reach of human eyes. Telescopes scattered across the globe allow us to explore the entirety of the night sky, and expeditions to study eclipses and occultations allow us to chase the exciting and serendipitous moments of astronomy. Smaller telescopes also allow us to keep exploring the countless mysteries of space that lurk in our own cosmic backyard, studying bright nearby stars that still hold the answers to innumerable puzzles. Gravitational waves hold the answers to questions about the universe that we've only just begun to ask.

In short, we *need* new and technologically brilliant telescopes like the Rubin Observatory—great strides in science have always gone hand in hand with the new capabilities offered to us by technology—but they won't do everything. One colleague compared the suite of telescopes that astronomers need for studying the universe to the full suite of appliances and tools used in a busy kitchen. A good cook may delight in the things they can do with a top-of-the-line KitchenAid, but to make a gourmet meal, they also need pots, pans, some simple bowls and spatulas, and they may even bust out their grandmother's old hand mixer once in a while.

Unfortunately, the next generation of telescopes is being built in an era when funding for astronomy is becoming increasingly tight. While funding is, so far, still being set aside to support cutting-edge new telescopes, budgets are being squeezed elsewhere to try and do this as cheaply as possible with ever-more-limited resources for scientific research. Smaller telescopes across the country are being closed due to lack of funding or to divert sparse funds to larger—and, in many cases, more automated—projects. Observatories have been told to pick their oldest and least-productive telescopes and shut them down, even when these telescopes are still frequently used by astronomers for exciting new science. These same funds also pay the salaries and support the research efforts of the astronomers who *use* the telescopes. The net result is that we're faced with an ever-shrinking amount of money to actually turn the coming onslaught of new data—invaluable but impenetrable heaps of zeroes and ones—into *science*, the exhilarating discoveries that make headlines and grace magazine covers and capture the human imagination.

The excitement and innovation of massive new telescopes is impossible to turn away from. Anyone who recalls shivering in a cold prime focus cage and laboring over another not-yet-developed plate can appreciate how impressive and efficient an endeavor like the Rubin Observatory will be. Doing in a night what used to take months, years, an entire career? Of course telescopes like this should be built.

The trouble comes when we're told that we can't have both, that smaller or more specialized telescopes must be shuttered to make way for the new behemoths. Many of them are described as old or obsolete rather than simply serving a different purpose. To return to the kitchen analogy, these older telescopes are chef's knives next to the next generation of telescopes' food processors: the latter can certainly cut more things than the former, but the two tools also do two very different jobs.

If funds for astronomical research continue to dwindle, the choice is clear. We need new telescopes, technological progress, and powerful automated or robotic telescopes that can crank out heaps of new data and open up new corners of the universe. The problem is that by doing this and *only* this, we're forced to sacrifice other aspects of the science. Survey telescopes will wildly expand our ability to find the unusual things in the night sky, but we also need to explore them. Studying millions of objects at a time may be efficient, but we still need to chase down the oddballs and rarities, the one strange outlier that could reveal surprising new physics or a signal from a distant world. There's research on the universe that simply can't be done with a robot or an automated pipeline, oddities in the sky that require a human eye to separate out the exciting from the errors, and discoveries that grow out of the ideas squeezed into the observing plan or the last few minutes of darkness at the end of a night. This type of science is at risk of disappearing if automated telescopes end up wholly replacing, rather than complementing, human-driven observing.

As the technology of astronomy moves forward, the way astronomers do our jobs is beginning to evolve too. Some changes are unarguably good: the volume of data and its broad availability are both wonderful when it comes to providing scientific resources for the next generation of astronomers. Operating a telescope remotely from your office is also certainly less physically strenuous than traveling to one in person. When using data from a robotic telescope, there are no platforms to fall off of, no scorpions or tarantulas scurrying around the control room, no arduous expeditions to Argentina or the South Pole or the stratosphere in pursuit of a few moments of observations.

However, we're also losing the experiences, the stories, the adventures of observing. To be clear, none of us are pining for the good old days or wishing we could turn the clock back to a time when we knew less about the universe and had fewer tools at our disposal to explore it. Still, the hands-on era of observing represents a type of scientific adventure that is starting to dwindle, and the excitement of it serves its own purpose.

o o o

Perhaps the most fundamental question about astronomy, one every astronomer is asked at some point in our careers, is why? Why do we do it? Why do we dedicate time and energy and resources to build and use telescopes for the sole purpose of studying things that are millions or billions of light-years away? The question comes from family members, from friends, from strangers on planes and trains, from those who control the purse strings for scientific research and those who simply want to understand what has possessed a small fraction of humanity to look away from our own planet and toward the gaping expanse of the universe and say, "I'm going to study *that*."

Usually this common why stems from one of three distinct questions. Why did we, as individuals, decide to study astronomy? Why is it worth the money? And why should *humanity* study astronomy?

There are several ready answers for why we, as humans, should study the universe. There's the purely practical perspective: developing new telescopes involves inventing new technology, and innovation always holds the potential for future tangible improvements in our day-to-day lives. Understanding strange new physics can open the door to new solutions for problems like efficient energy and transportation. Astronomy certainly could, I suppose, lead to down-to-earth everyday gains.

Astronomy also serves a much more powerful purpose as an excellent gateway science. It may not have the day-to-day applications of medical research or engineering, but the universe is excellent at capturing the human imagination. Teaching someone to love the universe, to be curious, to ask a seemingly silly question and then hunt doggedly through the math and physics that holds the answer is a way to spark scientific curiosity that can manifest in a thousand different ways in a young scientist. Someone may read about black holes and want to devote their life to black holes, or they may get curious about how the computers that study black holes work and want to build a better computer, or they may think *black hole* is a ridiculous name and decide to go take over whatever committee of goofs is in charge of naming the stuff we study in the heavens. The sheer beauty and scale of astronomy can propel someone into the world of science and research.

I also always get a kick out of imaging some more fantastical responses to these questions. When asked why we need to study things like the insides of stars or how our solar system works, an odd but enjoyable answer is: aliens.

It goes without saying that discovering or contacting intelligent extra-terrestrial life would be a paradigm-shifting moment for humanity and an incredible astronomical achievement in its own right. Still, I like to imag-ine what comes next. Imagine that we've contacted aliens on some distant planet and somehow developed the ability to communicate. What would we say to them? What do you talk about when you meet someone new? You talk about the immediate things you have in common: the weather, the news. The early conversation is "Gosh, it is just *pouring* out today!" or "Oh, have you heard about the latest news story?" We have no idea what we'll have in common with an alien intelligence…except for the universe. Instead of rain or breaking news, they might begin the conversation with "Gosh, that meteor shower was just coming down cats and dogs on Earth a little while back; how'd that work out for your dinosaurs?" or "Oh, did you see that exploding star?!" Our understanding of astronomy gives us the ability to be good conversationalists, to share what we've learned about the galaxy around us. Eventually, we may move on to other forms of expres-sion, to sharing who we are and our dreams and souls, but the universe and the language of science is the one common ground and starting point we know we'll have.

Of course, regardless of how outlandish or practical the answer is, some people will press further, wanting to dig into the monetary and motivational side of the problem. Stars are neat and all, the attitude seems to be, but what are they *worth*? What does astronomy return to us in dollars and cents and tangible benefits? Why should we spend time and money on the universe instead of something else? In a time, the argument goes, when there are so many complex and serious problems facing us here on our own planet, why would a group of researchers, however small, turn their attention and funding away from everything we run across in everyday life? At times, there's an unspoken accusation in this question: why are we devoting resources to space when we could be curing cancer or stopping climate change?

This apparent choice—between astronomy or some other area of research—is one of the current banes of scientific progress. Astronomers share this quandary with almost every other field of scientific research, constantly told there aren't enough resources available for what we're trying to do: not enough to fund both a large survey telescope *and* a large mirror telescope, not enough to study stars *and* cure diseases or protect the environment, not enough for the universe. The reality is that these are choices that shouldn't have to be made. It's true that there isn't an infinite wellspring of funding to draw from, but at the current scale of scientific research, even a few more droplets for science as a whole would make, literally, a universe of difference.

Finally, when it comes to why we, as people, as individuals, as astronomers, study the sky, the answers are a bit different.

In all the time I spent interviewing friends and colleagues for this book, I never once asked anyone how they got interested in astronomy. I wasn't writing about our origin stories; I was interested in the quirks and hijinks and wacky stories that come from the odd type of work we do.

The vast majority of astronomers that I spoke to told me anyway. Many of us are simply accustomed to giving these answers, explaining how we wound up in such a small and unusual profession. It made sense: the tellings were usually stories, and observing stories at that. Most astronomers I spoke to were inspired to study the stars during nights spent shivering at a telescope and looking at the sky, and many people's most vivid memories of observing are the breathtaking moments of standing outside an observatory and staring straight up, off the surface of the earth and into the stars.

Still, even these tales don't quite get at the heart of the matter. Most people told me the nuts and bolts of their stories: standing at this telescope or that mountaintop, looking up on a clear night or peering through an eyepiece for the first time, and being shocked breathless as some immutable internal shift took place. The story would then end; they'd found the universe, and that was it. Translating this sort of story into an answer when someone asks me "Why do you, personally, do astronomy?" is always a difficult proposition. I know the person asking doesn't want to hear "Because space is cool!" *Everyone* agrees that space is cool. Still, far from everyone winds up studying it professionally. I think dinosaurs are cool too, but I'm not a professional paleontologist.

In truth, the reason why we study astronomy is so deeply engrained that it almost makes the question seem strange. I'd compare it to someone asking why you married your spouse. Answers like "I love them" or "We just…click" or "Well, we met and it just…worked and…so one day…" tend to come out, semicoherent and baldly incomplete.

For me, the best explanation—short and perhaps insufficient but also deeply accurate—is encapsulated in a brief exchange from the movie *The Red Shoes*. The protagonist, a passionate aspiring ballet dancer, is asked by someone, "Why do you want to dance?"

Her reply comes after only a moment's pause. "Why do you want to live?"

"Well…I don't know exactly why…but I must."

"That's my answer too."

It comes from inside. Some tiny intangible but un-put-outable fire, within some tiny creatures on a tiny planet, drives us to reach outward and upward into the vast cosmos before us simply because we *must*.

Why do we study the universe? Why do we look at the sky and ask questions, build telescopes, travel to the very limits of our planet to answer them? Why do we stargaze?

We don't know exactly why, but we must.

o o o

I wrote this book to capture the human stories of working at telescopes. The past decades of astronomy may have gathered less data than the telescopes of the future, but they also offered a rich set of experiences for the observers who lived through them. These astronomers' stories are wonderful, and their telling and retelling reveals a unique slice of the human endeavor of science, but they also represent an era we will likely not be getting back.

That said, this book is not meant to be a paean for the "good old days" or a lament on how technology is changing the world. The Rubin Observatory and the generation of telescopes that follow it will have stories of their own. There's a Rube Goldberg–like magic to having so many minds and hands and years of collective research folded into the automation of these groundbreaking telescopes. I look forward to reading the successor to this book, written thirty years in the future by someone who's most likely still in grade school right now, sharing wild tales about the challenges of wrangling unfathomable quantities of data, the quirks of the new telescopes they'll be using, and the incredible discoveries that come pouring out of them. Whether we're strapped to a prime focus cage or downloading data from our kitchen tables, the study of astronomy will still continue, feeding our curiosity and our humanity as we explore the universe.

READING GROUP GUIDE

1. What memories do you have of stargazing and the night sky?

2. What surprised you the most about what life as an astronomer is like?

3. In Chapter 3, Levesque describes a frustrating evening at work and how something as minor as weather can potentially impact her professional plans and personal life. Have you had a similar experience? How did you approach it?

4. How has your perception of scientists changed after reading this book?

5. Many people tend to see *science* and *art* as very different pursuits, but Levesque writes about her colleagues' love of music, their cartoons depicting life at observatories, their poetic descriptions of solar eclipses, and their appreciation for the beauty of the universe. Why do you think we see science and art as distinct? Has your opinion on this changed?

6. What scientific aspect of astronomy—the discoveries, the mysteries, the science behind how telescopes work—did you find most interesting? What did you learn the most about?

7. What new questions do you have about astronomy and astronomers after reading this book?

Katy Garmany

Doug Geisler

John Glaspey

Nathan Goldbaum

Bob Goodrich

Candace Gray

Richard Green

Elizabeth Griffin

Ted Gull

Shadia Habbal

Ryan Hamilton

Suzanne Hawley

JJ Hermes

Jennifer Hoffman

Andy Howell

Deidre Hunter

Zeljko Ivezic

Rob Jedicke

David Jewitt

John Johnson

Dick Joyce

Mansi Kasliwal

William Keel

Megan Kiminiki

Tom Kinman

Bob Kirshner

Karen Knierman

Kevin Krisciunas

Rolf Kudritzki

Briley Lewis

Jamie Lomax

Julie Lutz

Roger Lynds

Peter Maksym

Jennifer Marshall

Joseph Masiero

Phil Massey

Cody Messick

Nidia Morrell

John Mulchaey

Joan Najita

Kathryn Neugent

Dara Norman

Knut Olsen

Carolyn Petersen

Erik Peterson

Emily Petroff

Phil Plait

George Preston

John Rayner

Joseph Ribaudo

Mike Rich

Noel Richardson

Gwen Rudie

Stuart Ryder

Abi Saha

Anneila Sargent

Steve Schechtman

Brian Schmidt

Francois Schweizer

Nick Scoville

Alice Shapley

INTERVIEWS

I owe a great debt of thanks to the 112 friends and colleagues I interviewed for this book. Every single person I spoke to added a wonderful voice and perspective to the chapters you've just read, and I'm grateful to each of them for their time and generosity in sharing their stories.

Helmut Abt	*Charles Danforth*
Bruce Balick	*Jim Davenport*
Eric Bellm	*Arjun Dey*
Edo Berger	*Trevor Dorn-Wallenstein*
Emily Bevins	*Alan Dressler*
Ann Boesgaard	*Maria Drout*
Howard Bond	*Oscar Duhalde*
Mike Brown	*Patrick Durrell*
Bobby Bus	*Erica Ellingson*
David Charbonneau	*Joseph Eggen*
Geoff Clayton	*Travis Fischer*
Andy Connolly	*Kevin France*
Thayne Currie	*Wes Fraser*

Bruno Sicardy

David Silva

Jeffrey Silverman

Josh Simon

Brian Skiff

Brianna Smart

Alessondra Springmann

Sumner Starrfield

Chuck Steidel

Woody Sullivan

Nick Suntzeff

Paula Szkody

Kim-Vy Tran

Sarah Tuttle

Patrick Vallely

George Wallerstein

Jonelle Walsh

Larry Wasserman

Jessica Werk

David Wilson

Charlotte Wood

Sidney Wolff

Jason Wright

Dennis Zaritsky

NOTES

1   George Wallerstein, interview with the author, August 9, 2017.

2   Richard Preston, *First Light: The Search for the Edge of the Universe* (New York: Random House, 1996), 263.

3   Michael Brown, interview with the author, July 24, 2018.

4   Sarah Tuttle, interview with the author, August 18, 2018.

5   Rudy Schild, "Struck by Lightning," 2019, http://www.rudyschild .com/lightning.html.

6   Geisler, Doug. 76 cm Telescope Observers Log—Manastash Ridge Observatory. Night log entry, May 18, 1980.

7   Howard Bond, phone interview with the author, December 6, 2018.

8   Greg Monk, quoted in "The Collapse," in *But It Was Fun: The First Forty Years of Radio Astronomy at Green Bank*, ed. F. J. Lockman, F. D. Ghigo, and D. S. Balser (Charleston: West Virginia Book Company, 2016), 240.

9   Harold Crist, quoted in "The Collapse," *But It Was Fun*, 241.

10 George Liptak, quoted in "The Collapse," *But It Was Fun*, 241.

11 Crist, quoted in "The Collapse," *But It Was Fun*, 243.

12 Ron Maddalena, quoted in "The Collapse," *But It Was Fun*, 245.

13 Pete Chestnut, quoted in "The Collapse," *But It Was Fun*, 247.

14 Anneila Sargent, interview with the author, July 2, 2018.

15 George Preston, interview with the author, June 5, 2018.

16 Harlan J. Smith, "Report on the 2.7-meter Reflector", Central Bureau for Astronomical Telegrams, Circular 2209 (1970): 1.

17 Marc Aaronson and E. W. Olszewski, "Dark Matter in Dwarf Galaxies," in *Large Scale Structures of the Universe: Proceedings of the 130th Symposium of the International Astronomical Union, Dedicated to the Memory of Marc A. Aaronson (1950–1987), Held in Balatonfured, Hungary, June 15–20, 1987*, ed. Jean Audouze, Marie-Christine Pelletan, and Sandor Szalay (Dordrecht: Kluwer Academic, 1988): 409–420.

18 University of Arizona Department of Astronomy, "Aaronson Lectureship," 2019, https://www.as.arizona.edu/aaronson_lectureship.

19 Elizabeth Griffin, interview with the author, January 8, 2019.

20 Anneila Sargent, interview with the author, July 2, 2018.

21 Vera C. Rubin, "An Interesting Voyage," *Annual Review of Astronomy and Astrophysics* 49, no. 1 (2011): 1–28.

22 Anne Marie Porter and Rachel Ivie, "Women in Physics and Astronomy, 2019," American Institute of Physics Report (College Park: AIP Statistical Research Center, 2019).

23 Porter and Ivie, "Women in Physics and Astronomy, 2019."

24 Leandra A. Swanner, "Mountains of Controversy: Narrative and the Making of Contested Landscapes in Postwar American Astronomy," PhD diss., Harvard University, 2013.

25 James Coates, "Endangered Squirrels Losing Arizona Fight," *Chicago Tribune*, June 18, 1990.

26 Thayne Currie, interview with the author, November 13, 2018.

27 John Johnson, interview with the author, March 28, 2019.

28 N. Bartel, M. I. Ratner, A. E. E. Rogers, I. I. Shapiro, R. J. Bonometti, N. L. Cohen, M. V. Gorenstein, J. M. Marcaide, and R. A. Preston, "VLBI Observations of 23 Hot Spots in the Starburst Galaxy M82", *The Astrophysical Journal* 323 (1987): 505–515.

29 D. Andrew Howell. Twitter Post. August 19, 2017, 1:43 a.m., https://twitter.com/d_a_howell/status/898782333884440577.

30 Oscar Duhalde, interview with the author, April 25, 2019.

31 Robert F. Wing, Manuel Peimbert, and Hyron Spinrad, "Potassium Flares," *Proceedings of the Astronomical Society of the Pacific* 79, no. 469 (1967): 351–362.

32 A. R. Hyland, E. E. Becklin, G. Neugebauer, and George Wallerstein, "Observations of the Infrared Object, VY Canis Majoris," *The Astrophysical Journal* 159 (1969): 619–628.

INDEX

ACKNOWLEDGMENTS

A massive thank-you is due, first and foremost, to Jeff Shreve. Jeff was there when the idea for this book was born, popping into my head without warning during our first meeting in a drafty exhibition hall. More than a year later, he was back on the scene as my literary agent to continue the process of bringing this book into the world. My deepest thanks to Jeff and the rest of the team at The Science Factory for being such wonderful champions of this book and so many other great scientific stories!

I'm immensely grateful to Anna Michels at Sourcebooks and Sam Carter at Oneworld for being such wonderful, insightful, and enthusiastic editors, and for guiding me through the process of honing a manuscript with so many moving parts into a final finished story. Thanks are also due to Grace Menary-Winefield for being one of this book's first advocates and champions; to Shana Drehs, Erin McClary, Chris Francis, Juliana Pars, Sabrina Baskey, Cassie Gutman, and everyone else who had a hand in crafting each and every one of these pages (and the beautiful cover they sit between!); and to Liz Kelsch, Lizzie Lewandowski, Michael Leali, Caitlin Lawler, Valerie Pierce, and Margaret Coffee for helping this book finds its way into readers' hands. Finally, my deep gratitude to Shirley Roberson and

Mary McHugh at Hughes Media Law Group for their excellent guidance in navigating the wild and new-to-me world of contracts and publishing.

This book owes its roots and its heart to the literally hundreds of colleagues that I've been fortunate enough to share science and stories with over the years. Particular thanks are due to the more than one hundred people that generously set aside hours of their time to meet with me in person or over the phone to tell their stories, as well as the many more than joined in the fun and piped up with answers and anecdotes on social media. Every single one of you has had a hand in shaping the contents of this book and I can only hope that I've done your contributions justice in bringing our quirky and unusual jobs to life!

The stories for this book were gathered from telescopes and observatories all over the planet, but I'd particularly like to thank the observatories that hosted me as a visitor while writing this book: Lowell Observatory, the National Optical Astronomy Observatory headquarters, and Kitt Peak National Observatory in Arizona; Mauna Kea Observatory in Hawaii; Las Campanas Observatory and the Vera C. Rubin Observatory in Chile; the Carnegie Observatories in California; the Laser Interferometer Gravitational-Wave Observatory in Hanford, Washington; and the NASA Stratospheric Observatory for Infrared Astronomy teams in Palmdale, California and Christchurch, New Zealand. It takes a tremendous number of energetic and dedicated people to run the observatories and research facilities of the world, and my colleagues and I owe a great deal to the staff at these and many other observatories for providing the crucial support needed to carry out a universe's worth of exciting science.

Special thanks are due to Jeff Rich for a fantastic and fascinating tour of Carnegie and my first visit to a plate archive! Jeff Hall and Catie Blazek gave me a place to stay and, as always, a delightful visit to Lowell Observatory. Katy Garmany, John Glaspey, and Dave Silva found me an office and heaps of great stories at NOAO. I owe Bo Reipurth and Thomas Greathouse my thanks for welcoming me into their busy workdays while visiting the Big Island of Hawaii. Joe Masiero was kind enough to let me crash his observing run and show me around Palomar Observatory, and was as always a wonderful source of stories and guidance. I'm immensely grateful to John Mulchaey, Leopoldo Infante, Javiera Rey, and Nidia Morrell for making my visit to Las Campanas possible, to Zeljko Ivezic, Ranpal Gill, and Jeff Kantor for bringing me to the summit of Cerro Pachon to visit the Vera Rubin Observatory, and to the many people working on the Rubin Observatory for taking the time to speak with me during my visit! Huge thanks to Amber Strunk for my wonderful visit to LIGO and to the staff at LIGO who showed me around and answered my many many questions. Nick Veronico, Kate Squires, and Beth Hagenauer made my first fascinating visit to SOFIA happen, and everyone on the crew helped bring the observatory to life for me that week even though we never made it into the air! A special thank-you is also owed to Jake Edmondson for the spare camera body that saved our skin and became the only reason we have any pictures from that trip. Finally, I can't thank Randolf Klein and Michael Gordon enough for the invite to fly on SOFIA as an observer and for the trip of a lifetime to the bottom of the world!

Thanks to the entire staff of Venture Coffee in Ballard for the dozens of macchiatos and 6:00AM opening times that made writing this book possible.

My colleagues in the Department of Astronomy at the University of Washington have been a crucial source of support and sanity as I've tried to balance the wacky dual lives of a tenure-track professor and first-time author. I'm especially grateful for the brilliant, hard-working, and frequently hilarious members of the UW Massive Stars research group—Jamie Lomax, Trevor Dorn-Wallenstein, Kathryn Neugent, Locke Patton, Aislynn Wallach, Brooke Dicenzo, Tzvetelina Dimitrova, Keyan Gootkin, Megan Kokoris, and Annie Shoemaker—for being endless fonts of enthusiasm and science as this book came to life.

Phil Massey took me on my first professional observing run and has been an invaluable mentor, colleague, and friend over the past sixteen years. I also had the great good fortune to first learn the ropes of astronomical observing from the incomparable Jim Elliot during his time at MIT. I've had many great mentors and collaborators over the years, but I feel truly lucky to have begun my astronomy career under these two excellent teachers.

The Levesque and Cabana clans, from Pépère on down through the newest and littlest members, have been a lifelong font of love, energy, encouragement, and gloriously loud and talkative family gatherings. To my brother Ben, thanks for bringing me outside for my very first observing trip; I still want to be like you when I grow up. Mom and Dad, thanks for reading pretty much everything I wrote from my first tragically spelled kindergarten imaginings up through an early excerpt from this book.

Sometimes love means chocolate and shoulder rubs. Sometimes love means coding a customized web app that can transcribe hundreds of hours of interviews and deciphering pages of legalese. My husband Dave somehow manages to seamlessly blend all of these forms of love and more on a daily basis, and I can most certainly say that writing this book would not have been the experience it was without him. Dave, I love you as big as the sky.

ABOUT THE AUTHOR

Emily Levesque is an astronomy professor at the University of Washington. Her research is focused on understanding how the most massive stars in the universe evolve and die. She has observed for upward of fifty nights on many of the planet's largest telescopes and flown over the Antarctic stratosphere in an experimental aircraft for her research. Her academic accolades include the 2014 Annie Jump Cannon Award, a 2017 Alfred P. Sloan fellowship, a 2019 Cottrell Scholar award, and the 2020 Newton Lacy Pierce Prize. She earned a bachelor's degree in physics from MIT and a PhD in astronomy from the University of Hawaii.

When she occasionally stumbles across some spare time, she attempts to spend it traveling, playing violin, skiing, messing with new recipes, or finishing triathlons very slowly. These plans are often waylaid by an old couch and a new book.

This is her first popular science book. She lives with her husband in Seattle.